Human-Like Decision Making and Control for Autonomous Driving

This book details cutting-edge research into human-like driving technology, utilising game theory to better suit a human and machine hybrid driving environment. Covering feature identification and modelling of human driving behaviours, the book explains how to design an algorithm for decision making and control of autonomous vehicles in complex scenarios.

Beginning with a review of current research in the field, **Human-Like Decision Making and Control for Autonomous Driving** uses this as a springboard from which to present a new theory of a human-like driving framework for autonomous vehicles. Chapters cover system models of decision making and control, driving safety, riding comfort and travel efficiency. Throughout the book, game theory is applied to human-like decision making, enabling the autonomous vehicle and the human driver interaction to be modelled using the noncooperative game theory approach. The text also uses game theory to model collaborative decision making between connected autonomous vehicles. This framework enables human-like decision making and control of autonomous vehicles, which leads to safer and more efficient driving in complicated traffic scenarios.

The book will be of interest to students and professionals alike, in the fields of automotive engineering, computer engineering and control engineering.

Human-Like Decision Making and Control for Autonomous Driving

Peng Hang
Chen Lv
Xinbo Chen

CRC Press
Taylor & Francis Group
Boca Raton London New York

CRC Press is an imprint of the
Taylor & Francis Group, an informa business

First edition published 2023
by CRC Press
6000 Broken Sound Parkway NW, Suite 300, Boca Raton, FL 33487-2742

and by CRC Press
4 Park Square, Milton Park, Abingdon, Oxon, OX14 4RN

CRC Press is an imprint of Taylor & Francis Group, LLC

Library of Congress Control Number: 2022938219

ISBN: 9781032262086 (hbk)
ISBN: 9781032262093 (pbk)
ISBN: 9781003287087 (ebk)

DOI: 10.1201/9781003287087

Typeset in CMR10 font
by KnowledgeWorks Global Ltd.

Publisher's note: This book has been prepared from camera-ready copy provided by the authors.

Contents

Preface

Human-Like Decision Making and Control for Autonomous Driving describes the latest human-like driving technology with game theoretic approaches. This book covers the feature identification and modeling of human-like driving behaviors, and the algorithm design of decision making and control for autonomous vehicles in various complex traffic scenarios, providing a better understanding of how the human-like driving framework will contribute to the decision-making system of autonomous vehicles in the human-machine hybrid driving environment.

This book starts with a literature review of human-like driving for autonomous vehicles. Then, our human-like driving framework for autonomous vehicles is proposed. Human-like driving feature identification and representation are presented first. Before the human-like algorithm design, some system models of decision making and control are built, covering driving safety, ride comfort, travel efficiency, etc.

Game theoretic approaches are applied to the human-like decision making of autonomous vehicles. The interaction behavior between the autonomous vehicle and the human driver is modeled with the noncooperative game theoretic approach. The collaborative decision making between connected autonomous vehicles are modeled with the cooperative game theoretic approach. The proposed framework can realize human-like decision making and control for autonomous vehicles, satisfying the personalized driving and riding demands. Besides, with the collaborative decision making and control of connected autonomous vehicles in complex traffic scenarios, the traffic efficiency and driving safety can be improved remarkably.

In general, this book features examples of using game theory to model human-like decision making for autonomous driving, and provides state-of-the-art approaches of decision making and control for human-like autonomous driving. In addition, the advanced decision-making implementation in connected autonomous vehicles under various driving situations are discussed. The book can be used as a reference for graduate students and researchers of automotive engineering, control engineering and transportation engineering in research areas such as autonomous driving, decision making, control theory, game theory, etc.

This book will remain up-to-date for the next 5 years and we will also provide a subsequent revised version of this book in coming years to continue to make it up-to-date and applicable for researchers and practitioners.

Finally, the authors would like to thank Dr. Chao Huang, Dr. Yang Xing, Dr. Zhongxu Hu and Dr. Sunan Huang for their assistance in the writing of this book.

Peng Hang
Chen Lv
Xinbo Chen

Authors

Peng Hang is a research fellow with RRIS, Nanyang Technological University, Singapore. His research interests include decision making, motion planning and control for connected autonomous vehicles. He earned a PhD from the School of Automotive Studies, Tongji University in 2019. He has been a visiting researcher with the Department of Electrical and Computer Engineering, National University of Singapore and a Software Engineer in the Research and Advanced Technology Department SAIC Motor, China. He has contributed more than 50 academic papers in the field of autonomous driving, and applied for more than 20 patents. He serves as an associate editor of *SAE International Journal of Vehicle Dynamics, Stability* and *NVH*, a section chair of the CAA International Conference on Vehicular Control and Intelligent, a guest editor of *Actuators*, *IET Intelligent Transport Systems*, *Autonomous Intelligent Systems* and *Journal of Advanced Transportation*, and an active reviewer for more than 40 international journals and conferences.

Chen Lv is currently an assistant professor at Nanyang Technology University, Singapore. He earned his PhD from the Department of Automotive Engineering, Tsinghua University, China in 2016. From 2014–2015, he was a joint PhD researcher in the EECS Dept., University of California, Berkeley. His research focuses on advanced vehicle control and intelligence, where he has contributed over 100 papers and obtained 12 granted patents. Dr. Lv serves as academic editor/editorial board member for *IEEE Transactions on Intelligent Transportation Systems*, *SAE International Journal of Electrified Vehicles*, etc., and guest editor for *IEEE/ASME Transactions on Mechatronics*, *IEEE Intelligent Transportation Systems Magazine*, etc. He received many selective awards and honors, including the Highly Commended Paper Award of IMechE UK in 2012, the NSK Outstanding Mechanical Engineering Paper Award in 2014, the CSAE Outstanding Paper Award in 2015, the Tsinghua University Outstanding Doctoral Thesis Award in 2016, the IV2018 Best Workshop/Special Issue Paper Award, and the INTERPRET Challenge of NeurIPS 2020 competition.

Xinbo Chen earned a BS in mechanical engineering from Zhejiang University, Hangzhou, China in 1982. He earned an MS in mechanical engineering from Tongji University, Shanghai, China in 1985 and a PhD in mechanical engineering from Tohoku University, Japan, in 1995. From 1988–1995, he was an assistant professor at the School of Mechanical Engineering, Tongji University. From 1996–2002, he was an associate professor at the School of Mechanical Engineering, Tongji University. Since 2002, he has been a professor with the School of Automotive Studies, Tongji University. He is the author of more than 200 articles and more than 70 patents. His research interests include dynamic control of omni-directional electric vehicles and design and control of active/semi-active suspension systems.

Chapter 1

Introduction

1.1 Overview of Human-Like Autonomous Driving

Decision making is a vital part of autonomous driving technology. According to the information provided by the environment perception module, proper driving behaviors are planned by the decision-making module and sent to the motion control module [79]. Therefore, decision making is usually regarded as the brain of autonomous vehicle (AV) and bridges environment perception and motion control. The driving performance of AVs is effectively affected by the decision-making module, including driving safety, ride comfort, travel efficiency and energy consumption [42].

It can be expected that human-driven vehicles and AVs will coexist on the roads in the coming decades. Driving safety will not be the only requirement for AVs, and how to interact with human-driven vehicles in complex traffic conditions is also vital and worthwhile studying. AVs are expected to behave similarly to human-driven vehicles. To this end, the behaviors and characteristics of human drivers should be considered in the autonomous driving design [60]. If AVs' driving behaviors are considered as human-like, it would be easier for human drivers to interact with surrounding AVs and predict their behaviors, particularly with regard to multi-vehicle cooperation.

In fact, different drivers present individual preferences on driving safety, ride comfort and travel efficiency, indicating their different

DOI: 10.1201/9781003287087-1

driving styles. For instance, in terms of collision avoidance in emergency conditions, aggressive drivers may choose a fast steering response, while timid drivers may choose to operate the braking pedal. Additionally, different passengers may also request diverse driving styles. For instance, if pregnant women, elderly people as well as children, are onboard, a more comfortable and safer riding experience is expected. In contrast, for daily commuters and those who are rushing to catch flights, travel efficiency takes a higher priority during their trips. Therefore, from the driving style perspective, the human-like decision making of AVs is expected to provide personalized options for passengers with diverse requirements on driving safety, ride comfort and travel efficiency.

1.2 Human-Like Decision Making for Autonomous Vehicles

To realize human-like decision making for AVs, different frameworks, methods, algorithms and strategies have been studied. In general, they can mainly be divided into three types, model-based human-like decision making, data-driven human-like decision making and game-theoretic human-like decision making [222].

1.2.1 Model-Based Decision Making

In recent years, some studies have introduced the human-like elements, typically represented by driving characteristics and driving styles, into the algorithm development for autonomous driving [138, 165]. Specifically, in terms of the decision-making functionality of AVs, the human-like model is expected to generate reasonable driving behaviors similar to our human drivers. To address the decision-making problem as well as the interactions with surrounding vehicles, a human-like behavior generation approach is proposed for AVs using the Markov Decision Process (MDP) approach with hybrid potential maps [64]. It is able to ensure driving safety generate efficient decisions as well as appropriate paths for AVs. Probabilistic model is widely used to deal with the uncertainties of decision making [195, 175]. In [268], a Bayesian networks model is applied to decision making that considers the uncertainties from environmental perception to decision making. However, it is difficult to deal with complex and dynamic decision missions. MDP is another common probabilistic model for decision making. In [9], a robust decision-making approach is studied using partially observable MDP considering uncertain measurements, which can make AVs move safely and efficiently amid pedestrians. In [189], a stochastically verifiable decision-making framework is designed for AVs based on Probabilistic Timed Programs (PTPs) and Probabilistic Computational Tree Logic (PCTL). In addition, the Multiple Criteria Decision Making (MCDM) method and the Multiple Attribute Decision Making (MADM) method are effective to deal with complex urban environment and make reasonable decisions for AVs [50, 21].

In a connected driving environment, AVs can share their motion states and thought easily, which is beneficial to safe and efficient decision making [172, 41]. To address the lane merging decision making for connected autonomous vehicles (CAVs), the centralized control approach has been widely studied by many researchers [185, 34]. A centralized controller is proposed to control and manage the vehicles within the certain controlled area [253]. Upon entering the controlled area, CAVs need to hand over their control authorities to the centralized controller. Then, the centralized control unit optimizes the entire traffic sequence and guides the vehicles on the on-ramp lane to enter the main lane orderly at a defined merging point.

To maximize the travel efficiency of CAVs, a longitudinal freeway merging control algorithm, in which CAVs follow the order assigned by the roadside controller, is proposed [124]. In [94], a broadcast controller is designed considering the pseudo perturbation to coordinate the CAVs on multiple lanes to realize smooth merging behaviors. To reduce the computing time and enhance the coordination performance, a grouping-based cooperative driving approach is studied to address the merging issue of CAVs [250]. In [89], a cooperative ramp merging framework is built for CAVs and human-driven vehicles with a bi-level optimization, which enables the cooperative and noncooperative behaviors in the mixed traffic environment. In addition, the vehicle platooning is an effective way to improve the merging efficiency [160]. In [196], a communication network is designed to control the lane-change actions of CAVs and the merging behaviors of the vehicle platoon. In [56], model predictive control (MPC) is applied to the vehicle platooning control to solve the multi-vehicle merging issue. Moreover, the concept of spring-mass-damper system is also utilized to improve the platoon's travel efficiency and stability after merging [10]. The aforementioned studies mainly focus on optimizing the merging sequence of CAVs (i.e., the longitudinal motion optimization) to advance the efficiency of traffic flow [23, 167]. However, the traffic scenario is mainly limited to a single-lane main road and an on-ramp lane, and the lane-change behavior of the vehicle on the main lane is usually neglected. Besides, the lateral motion optimization of CAVs during merging is rarely studied. To improve the merging efficiency on a multi-lane road, a MPC-based framework is established, which is able to generate the optimal acceleration and make safe lane-change decisions simultaneously [109].

To deal with decision making of CAVs at intersections, the concept of centralized traffic management system, in which a centralized controller is proposed to manage CAVs within the controlled area, has been widely studied [139, 253, 232]. In such cases, when a vehicle enters the controlled area, it must hand over the control authority to the centralized controller. Then, the centralized controller guides the vehicle to pass through the intersection according to an optimized passing sequence [231]. In [208], a distributed coordination algorithm is developed for dynamic speed optimization of CAVs in the urban street networks for improving the efficiency of network operations. However, the turning behavior of CAV is not considered in the traffic networks. In [248], a distributed conflict-free cooperation approach is proposed to optimize the traffic sequence of multiple CAVs at the unsignalized intersection for managing CAV operations with high efficiency, but it also requires a large amount of computation. To make a good trade-off between the system performance and computation complexity, a tree representation approach, which is combined with the Monte Carlo tree search and some heuristic rules, is proposed to find the global-optimal passing sequence of CAVs [251]. The aforementioned studies mainly focus on global optimization of the passing sequence and velocity coordination for CAVs at intersections. Although the centralized management system can improve the traffic efficiency, the computation burden largely increases with the rise of the amount of vehicles, which brings challenges in reliability and robustness [148]. However, if there is no global coordination at the intersection, then it is also a challenge to individual vehicles during their interactions with others. Therefore, the individual decision-making ability is quite important for CAVs [183]. To address the dynamic decision-making issue of CAVs at a roundabout intersection, a Swarm Intelligence (SI)-based algorithm is designed. In this algorithm, each CAV is regarded as an artificial ant that can self-calculate to make reasonable decisions with internet of things (IoT) technique [17]. To improve the reliability, robustness, safety and efficiency of CAVs at intersections, a digital map is used to predict the paths of surrounding vehicles, and afterward, the potential threats assessed are provided to the ego vehicle to make safe and efficient decisions [175]. In [30], a cooperative yielding maneuver planner is designed, which allows CAVs circulating inside single-lane roundabouts to create feasible merging gaps for oncoming vehicles.

1.2.2 Data-Driven Decision Making

With the development of machine learning algorithms, data-driven learning-based decision-making methods are gaining popular, including Support Vector Machine (SVM), Clustered SVM (CSVM), Extreme learning machine (ELM), Kernel-based Extreme Learning Machine (KELM), reinforcement learning (RL), Deep Neural Networks (DNN), etc. [81, 238, 93].

By learning driving behaviors and features from human drivers' driving data, data-driven approaches are easily to realize human-like driving and decision making. In [240], a KELM modeling method is proposed for speed decision making of AVs. Since the machine learning algorithm is developed based on probabilistic inference rather than causal inference, as a result, it is difficult to understand and find the failure cause of the algorithm. In [145], an SVM algorithm is applied to automatic decision making with Bayesian parameters optimization, which can deal with complex traffic scenarios. A novel decision-making system is built based on DNNs in [131], which can adapt to real-life road conditions. However, training of the DNNs requires a large number of data samples. In [59], combining a deep autoencoder network and the XGBoost algorithm, a novel lane-change decision model is proposed, which can make human-like decisions for AVs. In [19], a multi-point turn decision-making framework is designed based on the combination of the real human driving data and the vehicle dynamics for human-like autonomous driving. Learning from human drivers strategies for handling complex situations with potential risks, a human-like decision-making algorithm is studied in [29]. Since RL can provide many benefits in solving complex, uncertain sequential decision problems, in [289], a decision-making system is developed by integrating MDP and RL. In [267], a stochastic MDP is used to model the interaction between an AV and the environment, and RL is then applied to decision making based on the reward function of MDP. Compared with DNNs and other learning algorithms, the RL method does not need large size of driving dataset, instead, it leverages a self-exploration mechanism that solves sequential decision-making problems via interacting with the environment [219]. A Q-learning (QL) decision-making method is proposed based on RL in [170]. In order to learn optimized policies for decision making of AVs on highways, a multi-objective approximate policy iteration (MO-API) algorithm is proposed, which uses data-driven feature representation for value and policy approximation so that better learning efficiency can be achieved [255]. For most data-driven approaches, the decision-making performance relies on mass driving data to cover all possible driving scenarios. In [40], a hierarchical RL method is designed for human-like decision making, which does not depend on a large amount of labelled driving data.

However, the learning efficiency and generalization ability of the QL need to be further improved. To sum up, the learning-based decision-making methods, whose performances are limited by the quality of the dataset, still require further improvement.

In addition, RL is an effective approach to address the decision-making issue of CAVs at intersections. In [22], several state-of-the-art model-free deep RL algorithms are implemented into the decision-making framework, which advances the robustness when dealing with complex urban scenarios. In [57], an adaptive RL-based decision-making algorithm is studied, and an underlying optimization-based trajectory generation module is designed to improve the effectiveness of the decision making. However, learning-based approaches are data-driven ones, thus their performances are affected by the quality of the dataset. Moreover, it is difficult to address the algorithm failure due to the poor interpretability of the data-driven approach.

1.2.3 Game Theoretic Decision Making

Game theory is another effective way to formulate human-like decision making for AVs with social interactions. A game theoretic lane-change model is built in [271], which can provide a human-like manner during lane change process of AVs. In [226], the interactions between the host vehicle and surrounding ones are captured in a formulated game, which is supposed to make an optimal lane-change decision to overtake, merge and avoid collisions. Moreover, Stackelberg game theory is adopted to solve the merging problem of AVs in [266]. Based on the combination of game theory and MPC, a multi-lane vehicle ordering method is proposed to decide the optimal time and acceleration of lane change by considering the mutual interaction between vehicles [179]. In [265], a 3-person Stackelberg game theoretic approach is applied to the driver behavior modeling in highway driving. In [223], the single-step dynamic game with incomplete information is used to deal with the autonomous lane-change decision making for AVs, vehicle safety, power performance and passenger comfort are all considered in the lane-change process. In [249], the game theoretic approach is combined with RL, i.e., a Nash-Q learning algorithm, to realize human-like decision making for AVs. In [181], Nash equilibrium is applied to the decision making of multiple AVs at roundabout, which can find a good balance between driving safety and travel efficiency for AVs.

Besides, game theoretic approach shows superiority and effectiveness in interaction modeling and decision making of CAVs [110]. In a connected vehicular environment, game theory is used to predict lane change behavior [209]. Considering the interactive behaviors and characteristics between human-driven vehicles, the designed decision-making system is expected to be human-like. In [1], a game theory-based mandatory lane-change model (AZHW model) is proposed for the traditional environment and can be extended for the connected environment, which can effectively capture mandatory lane-change decisions with a high degree of accuracy. In [213], a decision-making algorithm is designed for CAV control at roundabout intersections with game theoretic approach. However, only two players, i.e., the ego vehicle and an opponent vehicle, are considered. In [212], the level-k game theory is applied to model multi-vehicle interactions at an unsignalized intersection, and this approach is combined with the receding-horizon optimization and imitation learning to design the decision-making framework for CAVs. In [104], a cooperative game approach is applied to the on-ramp merging control problem of CAVs, which can reduce the fuel consumption and travel time and further improve the ride comfort.

To explain that game theoretic approaches can realize human-like decision making and interaction, a simple instance is described as follows.

Figure 1.1 shows a typical lane-change decision-making scenario that includes four vehicles. If the ahead vehicle has lower moving speed than the host vehicle, the host vehicle will consider the corresponding decisions. Two strategies are provided for the host vehicle, i.e., slowing down and following the ahead vehicle, or changing lanes to the left lane. If the host vehicle chooses the second one, it must interact with the obstacle vehicle. For the obstacle vehicle, it can slow down and give ways for the host vehicle. Additionally, it can also speed up and fight for the right of way. Actually, the lane-change process of the host vehicle is a game between the host vehicle and the obstacle vehicle. The decision-making process is related to many factors, including the vehicle velocity, relative distance, and driving styles of vehicles.

Next, we will apply the Stackelberg game theoretic approach to the lane-change decision making of vehicles. In the Stackelberg game, three critical elements are introduced firstly, i.e., player, action and cost. In the lane-change decision-making issue, all vehicles are players. The action for the host vehicle can be simplified to two behaviors, i.e., lane-change and lane-keeping. The action for the obstacle vehicle can also simplified to two behaviors, i.e., acceleration and deceleration. The decision-making cost values are related to many factors. In this example, the costs of the four possible interaction cases are assumed to be fixed values shown in Table 1.1. For instance, if the host vehicle chooses lane-change and the obstacle vehicle chooses acceleration, the decision-making cost values for the host vehicle and the obstacle vehicle are 3 and 7, respectively. Correspondingly, if the host vehicle chooses lane-change and the obstacle vehicle chooses deceleration, the decision-making cost values for the host vehicle and the obstacle vehicle are 1 and 4, respectively. It can be found that both the host vehicle and the obstacle vehicle have smaller decision-making cost values in the second condition. Therefore, the second choices are better for both the host vehicle and the obstacle vehicle. Besides, if the host vehicle chooses lane-keeping and the obstacle vehicle chooses acceleration, the decision-making cost values for the host vehicle and the obstacle vehicle are 5 and 6, respectively. However, if the host vehicle chooses lane-keeping and the obstacle vehicle chooses deceleration, the decision-making cost values for the host vehicle and the obstacle vehicle are 4 and 5, respectively. Both the host vehicle and the obstacle vehicle have smaller decision-making cost values in the later condition. Therefore, if the host vehicle chooses lane-keeping, the obstacle will prefer deceleration.

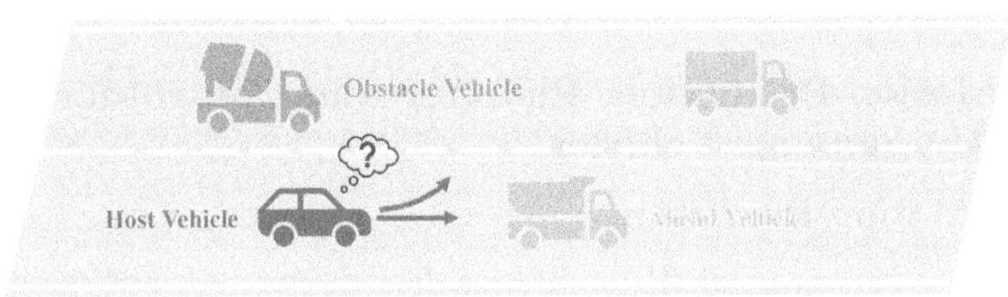

FIGURE 1.1: Schematic diagram of the lane change decision making for AVs.

In the Stackelberg game, both the host vehicle and the obstacle vehicle try to minimize the cost values. However, there exist a leader and a follower. The leader makes the decision firstly, and the follower makes the decision later. In the lane-change decision-making issue, the host vehicle is assumed to be the leader, and the obstacle vehicle is the follower. With the Stackelberg

TABLE 1.1: Action Costs with the Stackelberg Game Theory

Host Vehicle	Obstacle Vehicle	
	Acceleration	Deceleration
Lane-change	(3, 7)	(1, 4)
Lane-keeping	(5, 6)	(4, 5)

game theory, we can analyze the four cases in Table 1.1 again. If the host vehicle chooses lane-change, the obstacle vehicle will try to minimize its cost value. Since the cost values for the acceleration and deceleration are 7 and 4, the obstacle vehicle would like to choose the smaller one, i.e., deceleration. The equilibrium solution is (1, 4). However, if the host vehicle chooses lane-keeping, the obstacle vehicle will try to minimize its cost value as well. Since the cost values for the acceleration and deceleration are 6 and 5, the obstacle vehicle would like to choose the smaller one, i.e., deceleration. The equilibrium solution is (4, 5). After analysis, the leader, i.e., the host vehicle will choose a smaller cost value between 1 and 4. As a result, the host vehicle will change lanes, and the final equilibrium solution is (1, 4), which is also called the Stackelberg equilibrium. It can be found that the leader should consider all the cases and finally make the optimal decision. The follower can only minimize its cost after getting the decision-making result of the leader.

In general, the game theoretic approaches can be applied not only to handle the decision making of CAVs but also to simulate the interactive behaviors of intelligent agents.

1.3 Motion Prediction, Planning and Control for Autonomous Vehicles

1.3.1 Motion Prediction

Accurate motion prediction of surrounding agents, e.g., pedestrians and vehicles, is in favor of safer decision making and motion planning for AVs [113]. In general, existing motion prediction approaches can be divided into two types, i.e., model-based approaches focusing on the kinematic or dynamic model of vehicles, and data-driven approaches concerning the use of big data to analyze the hidden patterns of vehicles' trajectory [46].

The common model-based motion prediction approaches include the Constant Velocity Model (CVM), Constant Acceleration Model (CAM), Constant Turn Rate and Velocity Model (CTRVM), Constant Turn Rate and Acceleration Model (CTRAM), Constant Steering Angle and Velocity Model

(CSAVM), Constant Steering Angle and Acceleration Model (CSAAM), Interacting Multiple Model (IMM), etc. [63, 293, 194, 114]. Based on the vehicle kinematic model, considering uncertainties, CTRAM is combined with Unscented Kalman Filter (UKF) to realize short-term motion prediction for AVs [242]. To realize trajectory planning and safety assessment for AVs, Monte Carlo simulation is used to predict the probabilistic occupancy of the object, and the CTRAM-based MPC is applied to the trajectory prediction and optimization [229]. In [99], the motion prediction module employs an IMM filter to infer the intention of targets, and the prediction of future motion state is realized by fusing the prediction results of each model of the IMM filter. In [280], an integrated motion prediction model is designed for risk assessment and motion planning of AVs, which combines CTRAM, IMM, and the maneuver-based motion prediction model considering multiple interactions. In [97], the uncertainty propagation of motion prediction is modeled with the process update of the Kalman filter, and the vehicle kinematic is applied to the self-adaptive motion prediction of AVs. To predict the future motion of human-driven vehicles, the recursive least square (RLS) method is applied to approximate the dynamic model of the human-driven vehicles, and the triggering time and physical constraints are considered in the predictor design so as to improve the prediction accuracy [281]. In [122], a filtering scheme is designed for interaction-aware motion prediction for AVs based on an Interacting Multiple Model Kalman Filter equipped with novel intention-based models.

With the development of machine learning, different data-driven learning-based approaches have been applied to the motion prediction of AVs, including Long Short-Term Memory (LSTM), Recurrent Neural Network (RNN), convolutional neural network (CNN), generative adversarial network (GAN), etc. [274, 44, 239, 48]. The LSTM approach is effective to realize long-term motion prediction for AVs [154]. In [284], an LSTM-based framework that combines intention prediction and trajectory prediction together is proposed, which can deal with the motion prediction for AVs at intersections. To deal with the issue that LSTM models tend to diverge far away from the ground truth, a two-stage framework, i.e., Long-Term Network (LTN), is proposed for the long-term trajectory prediction of AVs, which has superior performance of long-term trajectory prediction than traditional LSTM approaches [230]. To improve the prediction of the LSTM algorithm, RRN is combined with LSTM, which shows good prediction performance for AVs at multi-lane turning intersections [98]. Considering the interaction between the ego vehicle and surrounding vehicles, an interaction-aware prediction algorithm is designed, in which the dynamics of vehicles are encoded by LSTMs with shared weights, and the interaction is extracted with a simple CNN [161]. In [180], a 3D CNN model is applied to the motion prediction to support the motion planning of AVs. In [287], a data-driven hybrid motion prediction model is designed for AVs based on LSTM and two 3D CNNs, which has better prediction performance than traditional deep-learning-based models. To address the propagating uncertainty in interaction behaviors, an attentive recurrent neural process (ARNP) approach is used for the motion prediction of AVs, which can predict lane-change trajectories in complex traffic scenarios [292]. With the combination of GNN, RNN and CNN, a Recurrent Convolutional and Graph Neural Networks (ReCoG) framework is applied to the accurate trajectory prediction of AVs [162]. With the Transformer-based Neural Network, a multi-modal motion prediction algorithm is designed, which can automatically identify different modes of the vehicles' attention and meanwhile improve the interpretability of the model [92]. Based on a GAN framework with a low-dimensional approximate semantic space, a novel motion prediction algorithm is designed for generating realistic and diverse vehicle trajectories [90]. To realize motion prediction in an interaction environment, a multi-modal hierarchical Inverse Reinforcement Learning (IRL) framework is designed to learn joint driving pattern-intention motion models from real-world interactive driving trajectory data, which can probabilistically predict the continuous motions partitioned by discrete driving styles and intentions [128]. Considering the interactions of AVs with surrounding agents and traffic infrastructures, a three-channel framework together with a novel Heterogeneous Edge-enhanced graph ATtention network (HEAT) is designed for the motion prediction of AVs, which can realize simultaneous trajectory predictions for multiple agents under complex traffic situations [163].

In general, model-based approaches show good prediction capability in short-term prediction with high computational efficiency, and data-driven approaches have a remarkable advantage on the long-term prediction [35]. According to different demands, such as prediction time, computational efficiency, driving scenario, etc., different motion prediction algorithms can be adopted for the motion planning and decision making of AVs.

1.3.2 Motion Planning

For AVs, the motion planning module is a bridge between the decision-making module and the motion control module [73]. Once the decision-making process is finished, the motion planning module will plan the acceptable velocity and path or trajectory for AVs. The main mission of the motion planning module is to realize collision avoidance. According to the planned velocity and path, the motion control module then conducts the velocity and path tracking control [76]. In general, the velocity planning is usually considered in the decision-making module. Therefore, this section mainly reviews path planning and trajectory planning. The planned path is only related to the position coordinates. Besides the position coordinates, the planned trajectory is also associated with time. Therefore, trajectory planning is usually conducted to address the collision avoidance of moving obstacles, e.g., emergency collision avoidance on highways. In general, path planning is usually used for global planning with static obstacles. With the collaboration with velocity planning, path planning can also deal with the collision avoidance of moving obstacles.

The planning algorithms can mainly be divided into five types: (1) graph algorithm, e.g., Voronoi diagram [247] and visibility graph method [277]; (2) heuristic search algorithm, e.g., A star (A^*) [140] and Dijkstra [201]; (3) random search algorithm, e.g., rapidly-exploring random trees (RRT) [36] and probabilistic road map (PRM) [258]; (4) potential field algorithm, e.g., stream function [234], simulated annealing [236], Laplace potential field [193] and boundary value problem (BVP) [45]; (5) spline curve, e.g., Dubins curve [68], Bessel curve [259], B-spline curve [13], sine curve [61], etc. To deal with the optimization issue of motion planning, intelligent optimization algorithms are usually adopted, including Genetic Algorithm (GA) [84], Ant Colony Optimization (ACO) [130], neural network [39], Particle Swarm Optimization (PSO) [150], etc. Though the intelligent optimization algorithms can help plan the optimal path, they rely on infinite iterations to approach a theoretical optimum, namely, the searching process may be too time-consuming due to computational load, which makes them hard to be applied to real-time online planning [129].

Graph algorithms are usually used for global path planning based on the graph that consists of nodes and edges as road map [261]. In the graph algorithm, the planned path is made up of a polygonal path, which is a piecewise linear curve and bends only at the obstacle vertices [243]. Visibility graph method is a typical graph algorithm for path planning, but it can only address the collision avoidance of static obstacles [202]. A^* algorithm is one of the most classical heuristic search algorithms [288]. To decrease the time consumption of A^* algorithm, D^* algorithm is proposed [5]. However, the quality of the planned path still depends on the density of the grid. Random search algorithms, e.g., RRT, can search and plan the optimal path with high efficiency. However, the planned paths have rough quality and need to improve [87]. Potential field algorithms are effective to realize path planning for AVs [164]. In the potential field algorithm, the vehicle is placed in a force field, and the planned path is produced by the effect of attraction and repulsion [192]. The main drawback of potential field algorithms is the local optimum issue. To address this issue, the improved BVP algorithm is proposed [119]. Besides, spline curves are usually used for local trajectory planning of AVs, e.g., lane-change, and overtaking [76]. The scenarios are relatively simple.

Based on the above literature review, it can be found that each planning algorithm has its limitation and drawback. To this end, multiple algorithms are usually combined together to advance the planning effect. In [129], a hybrid path planning algorithm is designed for AVs, which uses the GA algorithm for global path planning and a rolling optimization method for local path planning. In [199], a new GA algorithm is combined with the probability map for global path planning. In [264], GA algorithm is applied to the wolf pack search (WPS) algorithm for path planning. In [86], a dynamic path planning method is proposed for AVs, which can realize collision avoidance of both static and moving obstacles. In [26], a graph-based search algorithm is applied to the path planning for AVs in unstructured environments, and the Pythagorean Hodograph cubic curve is used to smooth the planned path. In [86], the artificial potential field approach and optimal control theory are utilized to design the path planner, which can plan the safe path for AVs and meanwhile guarantee the lateral stability. To plan the smooth and safe path for AVs, a dynamic path planner is designed with the polynomial parametric method and the GA algorithm [256]. To realize collision avoidance of moving obstacles in the overtaking process, an optimal planner is studied with the two-phase optimal control and the radio pseudo-spectral method [158].

To realize human-like motion planning, human-like driving features are considered in the aforementioned planning algorithms [65, 7, 143]. In [278], the human driver's characteristics are considered in the MPC trajectory planner. In [262], both the personalized factors of human drivers and the traffic environmental factors are considered to derive a personalized human-like lane-change trajectory planning model, in which personalized parameters of the longitudinal and lateral models are calibrated by the driving data with multi-dimensional time series regression method. In [58], a human-like path planner is designed based on a parameterized speed model fitted by human driving data, which can generate a reference path with smooth and peak-value-reduced curvature. In [191], RRT* is used to plan a human-like path for AVs. To generate human-like driving trajectory, a data-driven trajectory model is developed using general regression neural network (GRNN) [125]. In [126], by characterizing and modeling human driving trajectories using real vehicle test data, a data-driven trajectory model is designed with the long short-term memory neural network (LSTM NN).

1.3.3 Motion Control

The motion control of AVs consists of the longitudinal motion control and the lateral motion control [4]. The longitudinal motion control is usually reflected by the velocity or acceleration tracking control, and the lateral motion control is usually reflected by the path or trajectory tracking control.

Compared with the lateral motion control, the control objective of longitudinal motion control is usually single, i.e., velocity or acceleration. Therefore, the longitudinal motion control system is usually a single input single output (SISO) system. The control algorithm for longitudinal motion control is not very complex, which is not a technical difficulty [49]. To realize the motion control for adaptive cruise, the fuzzy logic approach is applied to the adaptive velocity tracking control [177]. In [282], the Backstepping control method is used for intelligent speed control, which comprehensively considers various road conditions including gradient, curve and weather. Considering the external disturbances and system nonlinearity, based on the Radial Basis Function (RBF) neural network, an adaptive sliding mode control (ASMC) algorithm is designed for velocity tracking control of AVs [74]. In [132], a parameter-varying controller is proposed based on the numerical reasonable model, which shows higher tracking accuracy and stronger robustness. Moreover, MPC is widely used for the longitudinal motion control. In [254], a receding-horizon MPC algorithm is designed for the longitudinal motion control of vehicles, which shows superiority to address the restricted nonlinear optimization. In [283], comprehensively considering the driving safety, velocity tracking, ride comfort and energy economy in the adaptive cruise control (ACC) system, MPC is used to realize multi-objective control.

At low-speed conditions, path tracking is usually an independent control mission for intelligent vehicles [52]. However, at high-speed conditions, the issue of vehicle dynamics and stability is unavoidable, which is usually comprehensively considered with path tracking [171]. Path tracking control aims to minimize the lateral offset and the heading angle error [83]. To realize emergency obstacle avoidance in high-speed conditions, a path tracking controller is designed combining 4WS feed-forward control and LQR feedback control [142]. However, the designed controller has poor robustness. To address the overtaking control issue, a four-wheel sliding mode control (SMC) driving controller and a four-wheel combined yaw rate and longitudinal velocity SMC driving controller are designed, which can effectively advance the active safety of vehicles [127]. In [216], a backstepping SMC method is applied to the path tracking controller design. However, the deigned controller is can only be used in low-speed conditions. With the comprehensive consideration of vehicle handling stability, safety and comfort, an integrated path tracking controller is designed with the feedforward and backstepping SMC, which can address the path tracking issue of vehicles in high-speed conditions [123]. Besides, robust control theory is effective to advance the robustness. In [82], $H\infty$ robust control method is applied to the integrated controller design for handling stability and path tracking, which can effectively maintain vehicle stability while suppressing lateral path tracking error in dynamic driving situations at handling limits. In [71], the μ synthesis robust control approach is used for the integrated control of path tracking and handling stability, which shows high control accuracy and strong robustness against parametric uncertainties. In [73], a LPV $H\infty$ robust controller is designed for path tracking, which can reduce the computational complex and guarantee the robustness.

Since the path tracking issue of AVs involves multiple objectives, MPC shows superiority over other control methods and has been widely used [28]. In [210], a novel path tracking algorithm is designed, which uses a kinematic MPC to handle the disturbances on road curvature, a PID feedback control of yaw rate to reject uncertainties and modeling errors, and a vehicle sideslip angle compensator to correct the kinematic model prediction. It shows good performance on steady-state and transient response, robustness and computing efficiency. In [237], a human-machine-cooperative-driving controller is designed with the collaborative control of the active front steering (AFS) technology and the direct yaw-moment control (DYC) technology, which utilizes MPC to advance the vehicle dynamic stability and path tracking performance. To realize the extreme path tracking control, e.g., small turning radius path, high speeds and on low adhesion roads, an integrated estimator of vehicle states and road adhesion coefficient is designed with an extended Kalman filter (EKF), and a four-wheel steering (4WS) path tracking controller is designed with MPC. As a result, the path tracking performance is improved remarkably [141]. Based on the driver model parameter identification, an active rear steering (ARS) MPC controller is designed to help drivers track the desired vehicle path, which can improve the driving safety and reduce the driver's workloads [276]. To realize the cooperative control of path tracking and handling stability, a linear-time-varying (LTV) MPC controller is designed based on the estimation of sideslip angles with the data fusion approach [66]. In [136], SMC is used for the longitudinal motion control, and LTV MPC is applied to the lateral motion control to predict the required front steering angle for path tracking.

To advance the control accuracy, three MPC path tracking controllers are designed, and the regime condition of a vehicle maneuver and the switching instant are determined by a fuzzy-logic-based condition classifier [204]. In [220], the fuzzy adaptive weight control is used to improve the path tracking performance of MPC controller, which shows better tracking accuracy and steering smoothness than conventional MPC controller. To deal with system nonlinearity, an NMPC method is applied to the ARS control in favor of the path tracking performance advancement [269]. To advance the robust performance against uncertainties and disturbances, e.g., road roughness and wind gusts, a robust MPC method is applied to the path tracking control [146]. In [80], a tube-based MPC approach is used for path tracking control of AVs, which comprehensively considers the control vector constraints, lateral stability constraints, rollover prevention constraints and path tracking error constraints. The integrated controller shows strong robustness in extreme conditions.

TABLE 1.2: Comparative Analysis of Different Control Algorithms

Control Algorithm	Accuracy	Robustness	Efficiency
PID	⋆	⋆	⋆⋆⋆⋆⋆
Fuzzy control	⋆⋆	⋆⋆	⋆⋆⋆⋆⋆
LQR	⋆⋆⋆	⋆⋆	⋆⋆⋆⋆
SMC	⋆⋆⋆⋆	⋆⋆⋆⋆	⋆⋆⋆
μ synthesis	⋆⋆⋆⋆	⋆⋆⋆⋆⋆	⋆⋆⋆
$H\infty$ control	⋆⋆⋆⋆	⋆⋆⋆⋆⋆	⋆⋆⋆
LPV $H\infty$ control	⋆⋆⋆⋆	⋆⋆⋆⋆⋆	⋆⋆⋆⋆
MPC	⋆⋆⋆⋆	⋆⋆⋆⋆	⋆⋆
NMPC	⋆⋆⋆⋆⋆	⋆⋆⋆⋆⋆	⋆

The comparative studies of different model-based control algorithms are described in Table 1.2. Three kinds of algorithm performances, i.e., accuracy, robustness and efficiency, are considered. More ⋆ means better performance. We can find that due to the simple structure and less control parameters, the control algorithms, e.g., PID and fuzzy control, have higher algorithm efficiency. However, the control accuracy and robustness are not satisfactory. Considering external disturbances and parametric uncertainties, robust control and SMC have higher control accuracy and stronger robustness, but the algorithm complexity is increased. With online optimization, MPC can also advance the control accuracy. However, the algorithm efficiency is decreased remarkably. In general, it is critical issue for controller design to find a good balance between control accuracy and algorithm efficiency.

Besides the above model-based control approaches, the data-driven approach or learning-based control, has been widely studied and used for the motion control of AVs in recent years [53, 153, 47]. In [105], the PID control parameters are self-tuning online with the self-learning and adaptive ability of a single neural network. To deal with the target-reaching and obstacle avoidance problem, a kind of recurrent fuzzy neural network is used, and the related learning method is designed to predict optimal control command [70]. In [198], three neural networks are used in the 4WS vehicle stability control system. In [103], an automatic lane-changing lateral control algorithm is designed for intelligent vehicles with a deep learning approach. To address the dynamic modeling and parameter identification problems in path tracking, a data-driven model-free adaptive control method is utilized, which does not depend on the kinematics and dynamics models of the vehicle and has high control accuracy [272]. In [102], deep deterministic policy gradient (DDPG) algorithm is applied to the control of LKA and ACC for intelligent vehicles, which has better active safety than traditional PID controller in extreme road conditions. Since neural network possesses good nonlinear approximation and learning ability, data-driven approaches are used to model the strong nonlinear system and combined with the model-based approaches to advance the control accuracy. In [115], reinforcement learning is used to decrease the path error by learning unknown parameters and updating a prediction model for MPC path tracking controller. In [101], an extreme learning machine is implemented to estimate the model-based predictive error for MPC controller, which can help reduce the path tracking error for intelligent vehicles. With the innovative Inverse Optimal Control (IOC) algorithm, it can learn a suitable cost function for the control task using collected data from human demonstration. As a result, the data-driven MPC controller is capable of learning the desired features of human driving and implementing them while generating the appropriate control actions [188].

To sum up, the control accuracy of the model-based control approach is associated with the accuracy of the control model and the algorithm complexity, which is a drawback of the model-based control approach. Complex control model and control algorithm can advance the control accuracy but will increase the computational complexity, which is harmful to the algorithm efficiency. Compared with the model-based control approach, the control accuracy of the data-driven approach does not rely on the accurate control model. As a result, data-driven approaches can be applied to various complex nonlinear systems. However, the control accuracy of the data-driven approach depends on the quality of the network and the amount of data. Therefore, a large amount of data should be collected for training. If the application scenario is not included in the training data, the data-driven approach may show bad control effect. Besides, the poor interpretability is another drawback of the data-driven approach. If there exists bad control effect, we cannot find the reason easily. The only solution is to retrain the model. For the model-based control approach, this issue can be easily addressed with controller parameter adjustment. In general, the data-driven approaches are usually used in an end-to-end autonomous driving that integrates decision making, motion planning and control rather than the single controller design.

1.4 Framework of Human-Like Autonomous Driving with Game Theoretic Approaches

Based on the literature review, we proposed a framework of human-like decision making and control for AVs, which is illustrated in Figure 1.2. It mainly consists of five modules. To realize human-like driving, the human-like driving features are defined for AVs firstly. Two kinds of approaches are given in this book. Driving aggressiveness is used to describe the aggressive

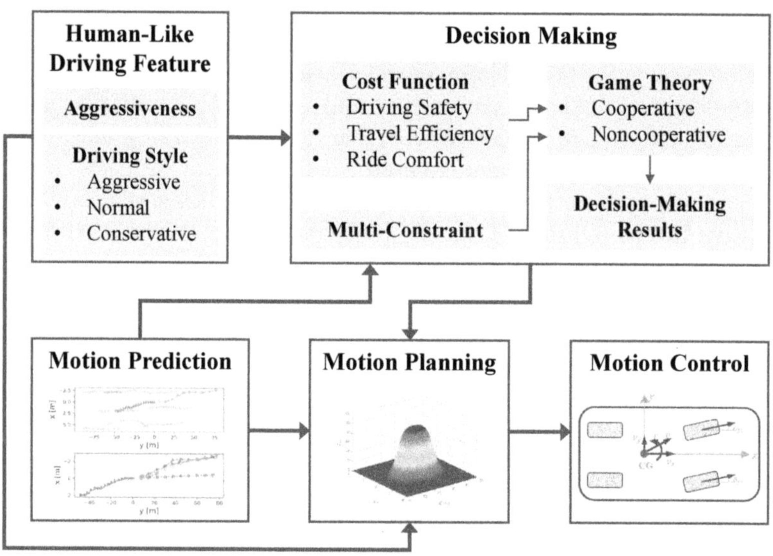

FIGURE 1.2: Framework of human-like decision making and control for AVs.

level of AVs, which is a continuous variable estimated according to the motion states, e.g., velocity, acceleration, etc. Another common method to describe the human-like driving feature is the driving style. In this book, the driving style of AVs is classified into three types, i.e., aggressive, normal and conservative. Different from the driving aggressiveness, the three driving styles are classified based on probability distributions. The results of human-like driving feature are outputted to the decision-making module and the motion planning nodule. In the decision-making module, the results of human-like driving feature are considered in the construction of the decision-making cost function related to driving safety, travel efficiency and ride comfort. Furthermore, multiple constraints are combined with the decision-making cost function, yielding the game theoretic decision-making issue. With the accurate prediction of motion states from surrounding vehicles, e.g., velocity, position and trajectory, the motion prediction module is in favor of the improvement of decision-making performance for the ego vehicle. According to different driving scenarios and objectives, the cooperative game theoretic approach and the nonoperative game theoretic approach are used to deal with the human-like decision-making issue of AVs. Then, the decision-making results are outputted to the motion planning module, in which driving styles and predicted motion states of surrounding obstacles will be considered to realize personalized and human-like motion planning. Finally, the planned trajectory and velocity will be sent to the motion control module. The tracking control of trajectory and velocity will be conducted.

Chapter 2

Human-Like Driving Feature Identification and Representation

2.1 Background

For human drivers, different drivers have distinguished driving styles, i.e., different drivers may make different decisions under the same condition. For instance, facing the overtaking behavior of an adjacent vehicle, aggressive drivers may choose to accelerate and stop the adjacent vehicle from overtaking. However, timid drivers may slightly decelerate and leave more space for others to pass through [147]. In recent years, the driving styles of human drivers have been widely studied [207]. Understanding driving style is in favor of the evaluation of vehicle performance, e.g., driving safety, travel efficiency, energy consumption, etc. [270, 190, 54]. The driving features of human drivers can be extracted for human-like driving of AVs [235].

Driving style is the driver's relatively stable, chronic and intrinsic behavioral tendency, and it is the synthesis of driver's mental thinking and behavioral patterns as well [203]. Driving styles are usually reflected by the drivers' behaviors, accelerating, decelerating and steering [227]. Physical representations are velocity, acceleration, yaw rate, etc. [156]. These physical signals are usually obtained by accelerometer, inertial sensor, and speed sensor, etc. Based on the obtained physical signals, different approaches are used to

DOI: 10.1201/9781003287087-2

identify or classify the driving styles, which will be introduced in the following section. The driving styles are usually divided into several represented categories, e.g., aggressive and nonaggressive. There is no unified standard or rule to define the driving style of aggressive or nonaggressive.

Driving style can represent the driving feature of human drivers. However, the classification of driving style is limited. Namely, human drivers are usually classified into several driving categories. The rough classification is too simple under some conditions, and cannot describe the detailed driving feature. Therefore, the driving aggressiveness index is proposed in some studies. The driving performances, such as driving safety, travel efficiency, and energy consumption, are considered in the design of driving aggressiveness index, which can represent the driving feature more accurately [214]. The identification approaches of driving aggressiveness are also discussed in the following section.

2.2 Driving Style Classification and Recognition

2.2.1 Classification of Driving Styles

In fact, different drivers present individual preferences on driving safety, ride comfort and travel efficiency, indicating their different driving styles. For instance, in terms of the collision avoidance in emergency conditions, aggressive drivers may choose a fast steering response, while timid drivers may choose to operate the braking pedal. Additionally, different passengers may also request diverse driving styles. For instance, if pregnant women, elderly people as well as children are onboard, a more comfortable and safer riding experience is expected.

In contrast, for daily commuters and those who are rushing to catch flights, travel efficiency takes a higher priority during their trips. In general, the driving style is defined as the manner in which the driver operates the vehicle controls within the driver context and driving conditions, which is related to many factors including gender, age, education and personality traits and personal characteristics such as self esteem, patience, recklessness, anger and extraversion [211, 157].

Different classification rules of driving style have been proposed. In general, two categories and three categories are two common classification rules. In [106] and [291], the driving style is characteristically divided into two types, i.e., non-aggressive and aggressive. In [186], the driving style is into two categories, i.e., aggressive or defensive. In [147], three different driving styles, i.e., aggressive, conservative and normal, are defined for AV.

a) Aggressive: This driving style gives the highest priority to the travel efficiency. Drivers would like to take some aggressive behaviors to reach their driving objectives including sudden accelerations or decelerations, frequent lane change or overtaking, and forcibly merging.

b) Conservative: It means cautious driving. Drivers care more about safety. Therefore, lane keeping, large gap and low speed driving are preferred.

c) Normal: Most drivers belong to this kind of driving style, which is positioned between the aforementioned two categories, expecting a balance among different driving objectives and performances [88].

2.2.2 Recognition Approaches for Driving Style

The driving style recognition algorithms can be mainly divided into the rule-based approach, the model-based approach, and the learning-based approach. The rule-based approaches include fuzzy logic, data threshold violation (DTV), etc. [107]. The model-based approaches include the decision tree [112], the Monte Carlo Markov model [62], and the Gaussian mixture model [227], etc. Recently, learning-based approaches, including neural network [252], Bayes learning [159], k-Means and support vector machine [218], etc., have become popular.

Fuzzy logic is an effective rule-based approach for driving style recognition [67]. In [37], fuzzy logic is applied to the online driving style recognition, which shows 68% correct classifications. The accuracy of the fuzzy logic approach depends of the amount of the combinatorial in fuzzy rules [2]. To overcome the combinatorial explosion from fuzzy rules, an improved online driving style recognition system is designed, which combines fuzzy logic with heuristics [38]. In [108], the accelerometer sensor data and the global positioning system (GPS) sensor deployed in smartphones are used to recognize driving styles with fuzzy logic. In [217], two driving styles, i.e., aggressive and normal, are proposed, and DTV is applied to the classification of driving styles based on the long-term accelerometer information. In [218], inertial sensors are used for driver classification and driving style recognition. Test results show that features associated with acceleration events do not play a significant role in differentiating between drivers, but features with braking and turning events showed significant potential in differentiating between drivers. In [18], drivers are classified based on the individual driver behavior, including the longitudinal and lateral control behaviors. The test dada of velocity, longitudinal and lateral accelerations are applied to the driving style identifier design. In [27], the comparative tests of three rule-based approaches, i.e., fuzzy logic, temporal logic paradigm, and Ar2p, are carried out. The test results indicate that temporal logic paradigm and Ar2p show better performance of driving style recognition.

In recent years, various model-based approaches have been studied for the driving style recognition as well [260]. In [20], a supervised hierarchical Bayesian model is proposed, in which the Labeled Latent Dirichlet Allocation (LLDA) is proposed to understand the latent driving styles from individual driving with driving behaviors, and the Safety Pilot Model Deployment

(SPMD) data are used to validate the performance of the proposed model. In [203], based on the root mean square of vehicle acceleration, the driving styles are defined and classified into three categories, i.e., steady, general, and radical, and Multi-dimension Gaussian Hidden Markov Process (MGHMP) is applied to the driving style recognition. In [197], based on energy spectral density (ESD) analysis and normalized driving behavior, an aggressiveness index was proposed to quantitatively evaluate the driving style. The proposed index is very suitable for those applications where the driving style is an important issue such as vehicle calibrations and intelligent transportation. In [206], a recursive algorithm under the Bayesian methodology is used for the driving style recognition, which considers the driving environment and fuel economy. In [69], the driver behavior uncertainty is considered in the driving style recognition algorithm based on the full Bayesian theory. Compared with the fuzzy logic method, it is more efficient and robust. In [182], two kinds of topic models are investigated: mLDA and mHLDA, to discover distinguishable driving style information with hidden structure from the real-world driving behavior data.

Besides, learning-based approaches, such clustering and shallow learning, are effective to realize driving style classification and identification [205, 178]. In [228], vehicle speed and throttle opening are treated as the feature parameters to reflect the two kinds of driving style, aggressive and moderate, and a k-means clustering-based support vector machine (kMC-SVM) method is applied to the driving style recognition. Based on the driving data from NGSIM, K-means and K-Nearest Neighbor (KNN) algorithm are respectively adopted in learning and recognition phases to recognize the current driving style pattern [135]. In [33], principal component analysis (PCA) and K-means clustering are utilized to classify 30 participants into cautious, moderate, and aggressive drivers. Based on three rear-end collision surrogates, Inversed Time to Collision (ITTC), Time-Headway (THW), and Modified Margin to Collision (MMTC), K-mean algorithm is used for training data labeling, and finally SVM is applied to recognize driving style based on the labelled data [257]. In [51], a driving style classifier is designed by supervised machine learning method, which can be used to recognize the driving style of drivers online. In [144], the driving data is collected, including the speed, acceleration, and opening degree of the accelerator pedal, and 44 feature quantities are extracted to characterize the driving style. Based on the feature quantities, the fuzzy c-means (FCM) clustering algorithm is combined with the SVM to identify the classified driving style. Learning from the smartphone sensors data, an ensemble learning method is used to classify the driving styles using SVM, multi-layer perceptron, and KNN [12]. Based on the physiological signals such as electroencephalography (EEG), a two-layer learning method is studied for driving behavior recognition with K-means, SVM, and KNN classifier [260]. In [246], three driving styles are proposed, i.e., conservative, calm and radical. Three kinds of clustering methods, i.e., K-means clustering, hierarchical clustering and PCA-based dimensionality reduction clustering, are applied to the recognition of driving styles. For driving style recognition, Recurrence Plot (RP) transforms sequential data into images and the converted images are processed into the driving styles by CNN, and the result of driving style recognition is used for trajectory prediction [25]. Based on a hybrid machine learning method that combines an unsupervised clustering method with a data-driven extreme learning machine (ELM) algorithm, the hierarchical clustering is applied to the driving style classification and identification, which comprehensively considers the driving safety, fuel efficiency and ride comfort [32]. In [173], unsupervised learning is used for driving style identification based on the database that includes 2736 drivers with 200 variable length driving trajectories each. Considering that existing clustering and shallow learning cannot accurately identify the types of abnormal driving behaviors, a recognition model is proposed based on a long short-term memory network and convolutional neural network (LSTM-CNN), which shows better performance than traditional clustering and shallow learning approaches [100].

Compared with the learning-based approaches, rule-based approaches and model-based approaches do not require large amount of data for training, and have higher computational efficiency. However, the accuracy of driving style recognition is poor than that of the learning-based approach. Online learning-based approaches have a large amount of computation for the hardware. Therefore, the balance between the computational efficiency and the accuracy is a critical issue for driving style recognition algorithm.

2.2.3 Characteristic Analysis of Different Driving Styles for Human-Like Driving

To realize the human-like driving and decision making for AVs, the driving behaviors of human drivers are analyzed based on the real-world driving dataset, i.e., the NGSIM dataset. The NGSIM vehicle data are collected from different regions in different time slots, reflecting different traffic scenarios [245]. In this section, two groups of driving data from NGSIM dataset, i.e., the I-80 and US-101 freeway sub-datasets, are applied to the driving behavior analysis of human drivers.

Driving safety, ride comfort and travel efficiency are three critical performances for driving behavior analysis. In this study, the time headway is used to reflect the driving safety. Vehicle acceleration is used to describe the ride comfort, and vehicle velocity is adopted to evaluate the travel efficiency. The mean values and the standard deviation (STD) values of the performance indexes are shown in Figure 2.1. It can be found that the aggressive driving style shows the largest travel velocity and acceleration among the three kinds of driving styles, indicating that aggressive drivers give more priority on travel efficiency rather than ride comfort and driving safety. Besides, the conservative driving style has the largest time headway and the smallest velocity and acceleration among the three kinds of driving styles, which means that the

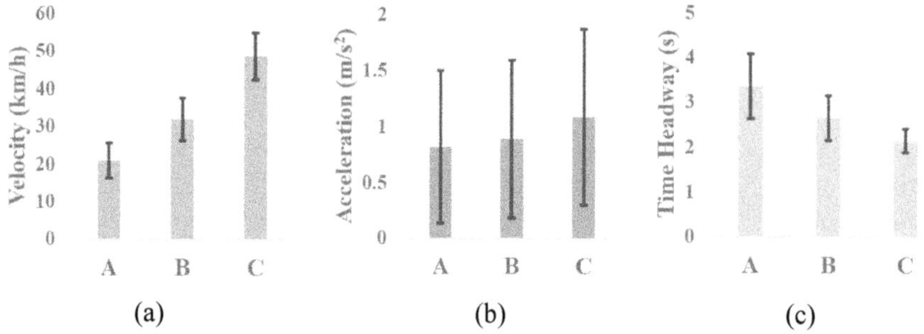

FIGURE 2.1: Characteristic analysis of different driving styles with the NGSIM data: (a) velocity; (b) acceleration; (c) time headway. The three driving styles are denoted by A: conservative, B: normal, and C: aggressive, respectively.

conservative driving style prefers better driving safety and ride comfort. The normal driving style is between the above two driving styles, finding a good balance between different driving performances. The above analysis can be applied to the human-like decision-making cost function design for AVs.

2.3 Driving Aggressiveness of Vehicles

2.3.1 Definition of Driving Aggressiveness

The is no uniform definition of driving aggressiveness. Many vague definitions of driving aggressiveness still exist. The aggressive driving is defined as a driving behavior that endangers or tends to endanger personal and property safety. Besides, the aggressive driving is defined as a driving behavior that is likely to increase the risk of collision or attempt to pursue high travel efficiency. The consensus is that the aggressive driving is related to the following driving behaviors: over-speed, frequent or sudden lane changes, abnormal acceleration and deceleration, inappropriate overtaking behavior, etc. [152]. From the above consensus, we can find that the driving aggressiveness is directly related to the longitudinal driving behavior, i.e., accelerating and decelerating, and the lateral driving behavior, i.e., steering. Some motion states of vehicles can be used to reflect the driving aggressiveness, e.g., longitudinal and lateral acceleration, velocity, yaw rate, sideslip angle, etc. In general, the driving aggressiveness can be obtained with subjective evaluation and objective evaluation. The subjective evaluation is to conduct a questionnaire-based survey on drivers. The objective evaluation is to conduct data analysis though some constructed indexes that can evaluate the driving aggressiveness level. The subjective evaluation approach usually reflects the subjective views of the driver rather than the actual performance of the driver on the road. Therefore, the data-based evaluation approach is widely used for driving aggressiveness evaluation.

2.3.2 Estimation Approaches of Driving Aggressiveness

The data-based estimation approaches of driving aggressiveness can mainly be divided into three types, i.e., the statistical regression approach, the time series analysis approach, and the machine learning approach.

The statistical regression approach and the time series analysis approach have been widely used for the estimation of driving aggressiveness [151]. Based on the Long Range (LoRa) communication networks, an aggressive driving detection system is designed with data analysis of vehicle motion states and operation information, in which risky driving behaviors such as rapid acceleration, rapid start, sudden deceleration, sudden stop and sudden turn are analyzed continuously. Besides, additional driving information such as idling time, mileage, travel time and top speed are used for aggressive driving detection [96]. In [134], the driving aggressiveness is assumed to be determined by focusing on four types of driving behaviors, i.e., steering, acceleration, deceleration, and alternation between acceleration and deceleration. The overall aggressiveness level is estimated with the overall aggressiveness score by integrating the four driving behaviors using the multiple liner regression. In [117], a driving aggressiveness detection approach is proposed using only visual information provided by forward camera, which is based on detection of the road lines and the vehicles on the road and extracts information related with road lane departure rate, speed of the vehicle and possible forward collision time. Using these extracted features, a classifier is applied to the driving aggressiveness estimation. In [120], by obtaining the information of driver' heart rate, steering wheel movements, and vehicle movement from the wearable electronic device, a classification model is designed for driving aggressiveness estimation. However, the primary limitation of this study is a relatively small sample dataset. In [116], two jerk metrics are designed to evaluate the driving aggressiveness, one for large positive jerk and the other for large negative jerk, when drivers are operating the gas and brake pedal, respectively. The study shows that the identification of aggressive driving could be reinforced by the number of large negative jerks, given that the drivers are tailgating, or by the number of large positive jerks, given that the drivers are categorized as violators. However, only the longitudinal driving behavior is considered in this study, the steering behavior is ignored. In [121], based on the data from in-vehicle sensors, a framework is designed to evaluate large-scale driving records and to establish clusters that can be used to identify potentially aggressive driving behaviors.

Besides, machine learning approaches are in favor of the driving aggressiveness estimation. To improve the overall traffic safety, the Iterative DBSCAN (I-DBSCAN) approach, an extension of the Density Based Spatial Clustering of Applications with Noise algorithm, is used to utilized as part of a machine learning analytic strategy for identifying aggressive driving behaviors within large, unlabeled RWD datasets [155]. In [166], a Long Short Term Memory Fully Convolutional Network (LTSM-FCN) is used to detect the driving aggressiveness, which outperforms the other methods in terms of the F-measure score. In a reinforcement learning based decision-making system (ReDS), a mixture density network based aggressive driving behavior detection method to is used to detect possible aggressive driving behaviors among surrounding vehicles, including sudden deceleration, sudden acceleration, sudden left or right lane change. Then, the aggressive driving behavior detection results are considered in the reward function for decision making [111]. In [95], an aggressiveness identification model is developed using the machine learning method of random forest to classify the value of time to lane crossing (TLC), a proxy for aggressive/risky driving, based on a set of motion related metrics as features. In [118], based on the fusion of visual and sensor features to characterize related driving session and to decide whether the session involves aggressive driving behavior, an SVM classifier is utilized to classify the feature vectors in order for aggressiveness decision.

In [152], based on the motion data collected by the accelerometer and gyroscope of a smart phone mounted on the vehicle. the comparative study of the Gaussian mixture model (GMM), partial least squares regression (PLSR), wavelet transformation, and support vector regression (SVR) are conducted. The empirical results show that GMM, PLSR, and SVR are promising methods for aggressive driving recognition. GMM and SVR outperform PLSR when only single-source dataset is used. PLSR performs the best when multi-source datasets are used. GMM and SVR are more robust to hyperparameter. In addition, incorporating multi-source datasets helps improve the accuracy of aggressive driving behavior recognition. In [55], the comparative study of three classes of algorithms, i.e., anomaly detection-bassed, threshold-based and machine learning-based, is conducted. The results show that machine learning-based approaches were able to achieve the best performance (especially SVM, followed by RF, CNN, and MLP), with higher computation time.

To realize human-like driving, the estimated results of driving aggressiveness is usually used in the algorithm design of human-like decision making and motion planning for AVs [6].

2.3.3 Aggressiveness Estimation Model for Human-Like Driving

To realize the driving aggressiveness estimation of vehicles, the driving behavior analysis of human drivers are conducted firstly based on the INTERACTION dataset [275]. The INTERACTION dataset contains naturalistic motions of various traffic participants. In this section, the motion data of human-driven vehicles at three merging and lane-change scenarios are analyzed. For human-driven vehicles, the longitudinal aggressiveness is usually reflected by the travel velocity, and the lateral aggressiveness is usually described by the yaw rate. Therefore, vehicle velocity and yaw rate are analyzed. The analysis results of the three scenarios are displayed in Figure 2.2.

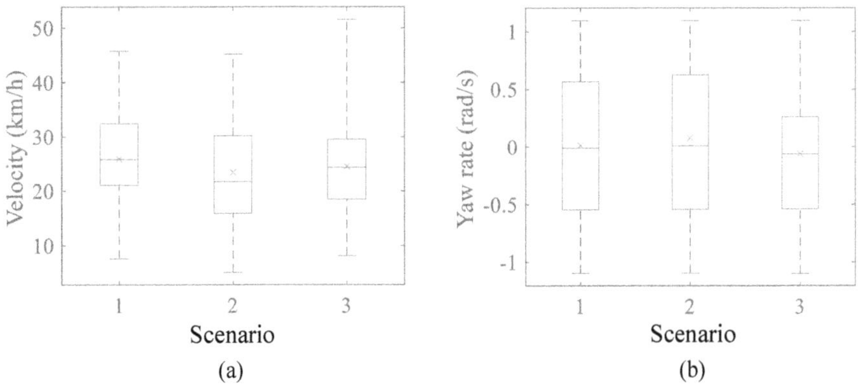

FIGURE 2.2: Driving behavior analysis of human drivers based on INTERACTION dataset: (a) velocity; (b) yaw rate.

TABLE 2.1: Fuzzy Logic for Aggressiveness Estimation

Velocity	Yaw rate				
	VS	S	M	L	VL
VS	C	C	C	N	N
S	C	C	N	N	A
M	C	N	N	A	A
L	N	N	A	A	A
VL	N	A	A	A	A

In this study, the fuzzy inference approach is applied to the driving aggressiveness estimation of vehicles. Vehicle velocity and yaw rate are the inputs of the fuzzy inference system, and the driving aggressiveness is the output. The fuzzy inference system consists of three critical components i.e., the fuzzification, the rule evaluation and the defuzzification [88]. The first step is fuzzification, which converts the continuous values of vehicle motion states into fuzzy values according to the membership functions. As Figure 2.3 shows, Z-shaped and triangular membership functions are used. Velocity and yaw rate are blurred into very small (VS), small (S), middle(M), large(L), and very large(VL). Additionally, the fuzzy values of aggressiveness are conservative

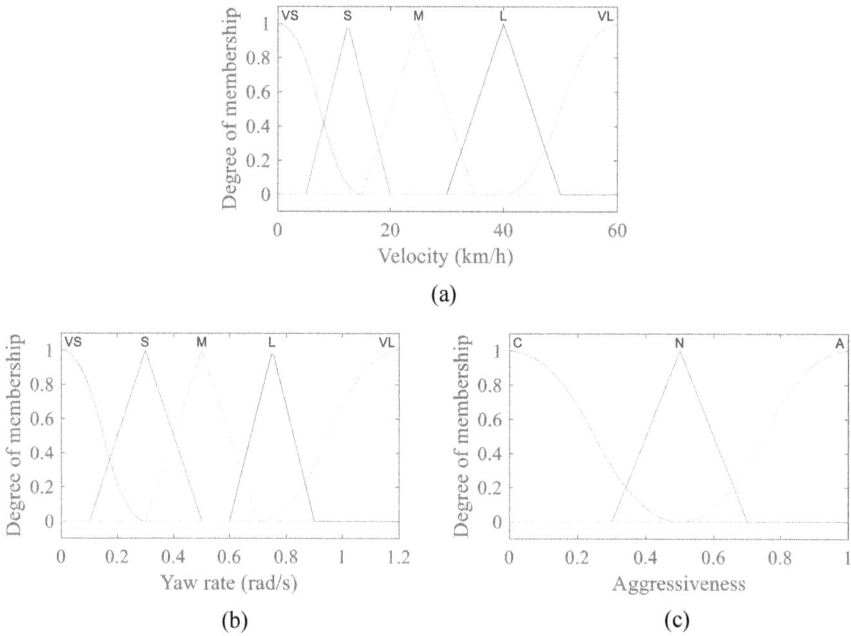

FIGURE 2.3: The membership functions: (a) velocity; (b) yaw rate; (c) aggressiveness.

(C), normal (N) and aggressive (A). After the fuzzification process of velocity and yaw rate is finished, the fuzzy value of aggressiveness can be inferred based on the fuzzy logic rule in Table 2.1. Once the fuzzy inference process is finished, the defuzzification will be conducted to obtain the real value of aggressiveness κ, $\kappa \in [0, 1]$. Finally, the driving aggressiveness estimation is done. The map of aggressiveness with respect to velocity and yaw rate is illustrated in Figure 2.4. We can find that the larger the velocity and yaw rate, the higher the aggressiveness. If the velocity or yaw rate exceeds the maximum, $\kappa = 1$.

2.4 Conclusion

In this chapter, to represent the driving feature, the concepts of driving style and driving aggressiveness are proposed. In general, the driving style are usually classified into several categories. The typical classification of driving style is aggressive, normal, and conservative. Three kinds of recognition

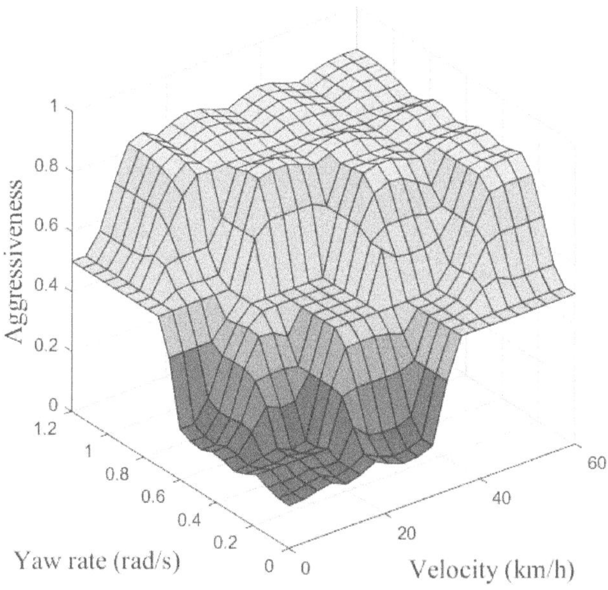

FIGURE 2.4: Aggressiveness MAP.

approaches for driving styles are introduced and discussed, including the rule-based approach, the model-based approach, and the learning-based approach. To analyze and extract the characteristics of human driving styles, a real-world driving dataset, i.e. the NGSIM, is used. The analysis and features extracted regarding the human driving styles are utilized in the cost function design of human-like decision making for AVs.

In addition to driving style, driving aggressiveness can represent the driving feature more accurately. Three typical estimation approaches of driving aggressiveness, i.e., the statistical regression approach, the time series analysis approach, and the machine learning approach, are studied and compared. Based on the driving behavior analysis of human drivers with the INTERACTION dataset, the fuzzy inference approach is applied to the aggressiveness estimation of surrounding vehicles. The estimated driving aggressiveness will be used in the human-like decision-making algorithm for AVs.

Chapter 3

System Modeling for Decision Making and Control of Autonomous Vehicles

3.1 Background

Different vehicles models have been proposed for decision making, motion planning, and motion control. In general, these vehicle models can be divided into two types, i.e., kinematic model and dynamic model. Kinematic models mainly describe the relationship between speed, acceleration, position and other motion states of vehicles, involving direction and size. Mass, moment of inertia, force, and torque are not involved in the kinematic model. To derive the kinematic model, the vehicle is usually abstracted as a mass point or a certain geometric shape. In this book, two kinds of vehicle kinematic model, i.e., mass point kinematic model and bicycle kinematic model, are proposed.

In general, vehicle kinematic models focus on the motion description of vehicles. In addition to the motion description of vehicles, the reason and source of vehicle motion are presented in the vehicle dynamic model. Therefore, force and torque are considered in the vehicle dynamic model. Therefore, besides the structure parameters of vehicles considered in the kinematic model, mass, moment of inertia, and other parameters associated with force and torque are required in the vehicle dynamic model. To this end, vehicle dynamic models are more complex than vehicle kinematic models.

DOI: 10.1201/9781003287087-3

Vehicle kinematic models and vehicle dynamic models are usually used in different conditions. Due to the simple structure and low computational load of vehicle kinematic models, vehicle kinematic models are usually applied to the decision making, motion prediction and motion planning algorithms. Moreover, it is also effective for low-speed motion control. Vehicle dynamic models are able to handle the issues of decision making, motion prediction and motion planning as well. However, it will increase the computational load. Therefore, vehicle dynamic models are usually used for the motion control algorithm design for AVs, especially for the high-speed motion control.

Driver models are effective to simulate the driving characteristics and behaviors of human drivers. With different parametric settings in the driver model, different driving styles can be described for AVs, yielding human-like driving for AVs.

3.2 Vehicle Model for Decision Making and Control

In this section, two kinds of vehicle kinematic model and two kinds of vehicle dynamic model are built for the algorithm design of decision making and control for AVs. Moreover, a driver model is established and combined with vehicle models to realize human-like driving for AVs.

3.2.1 Vehicle Kinematic Model

3.2.1.1 Mass Point Kinematic Model

The mass point kinematic model is usually used for path planning. In this model, AV is abstracted as a mass pint. As shown in Figure 3.1, a particle P is proposed, which runs along the path with a constant velocity from the start point S to the final point F, and the motion of the particle is conducted in an inertial frame described by the $X - Y$ coordinates.

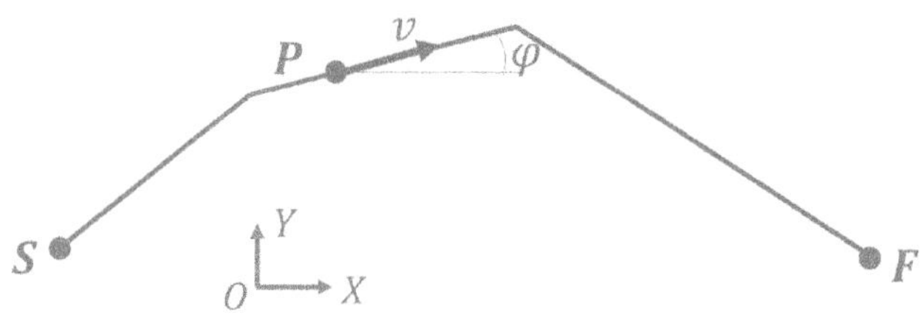

FIGURE 3.1: Mass point kinematic model.

The mathematical description of the mass point kinematic model is given by

$$\begin{cases} \dot{\varphi} = a_y/v \\ \dot{X} = v\cos\varphi \\ \dot{Y} = v\sin\varphi \end{cases} \tag{3.1}$$

where v and a_y are the velocity and lateral acceleration of the particle respectively, φ is the yaw angle of the particle, X and Y are the coordinate position of the particle.

Furthermore, (3.1) can be rewritten in the state-space form.

$$\begin{cases} \dot{x} = f[x(t), u(t)] \\ \dot{y} = g[x(t), u(t)] \end{cases} \tag{3.2}$$

where the state vector $x = [\varphi, X, Y]^T$, control input vector $u = a_y$, measurement output vector $y = [X, Y]^T$ and

$$f[x(t), u(t)] = \begin{bmatrix} a_y/v \\ v\cos\varphi \\ v\sin\varphi \end{bmatrix} \tag{3.3}$$

$$g[x(t), u(t)] = \begin{bmatrix} 0 & 1 & 0 \\ 0 & 0 & 1 \end{bmatrix} x(t) \tag{3.4}$$

The proposed mass point kinematic model is adopted for AV's path planning.

3.2.1.2 Bicycle Kinematic Model

To reduce the computational complexity, vehicle kinematic models have been widely used in the decision-making algorithm design [75]. In the bicycle kinematic model, AV is abstracted as a two-wheel bicycle. As Figure 3.2 shows, the bicycle kinematic model is derived as follows.

$$\dot{x}(t) = f(x(t), u(t)) \tag{3.5}$$

$$f(x(t), u(t)) = \begin{bmatrix} a_x \\ v_x \tan\beta/l_r \\ v_x\cos\Phi/\cos\beta \\ v_x\sin\Phi/\cos\beta \end{bmatrix} \tag{3.6}$$

$$\beta = \arctan(l_r/(l_f + l_r)\tan\delta_f) \tag{3.7}$$

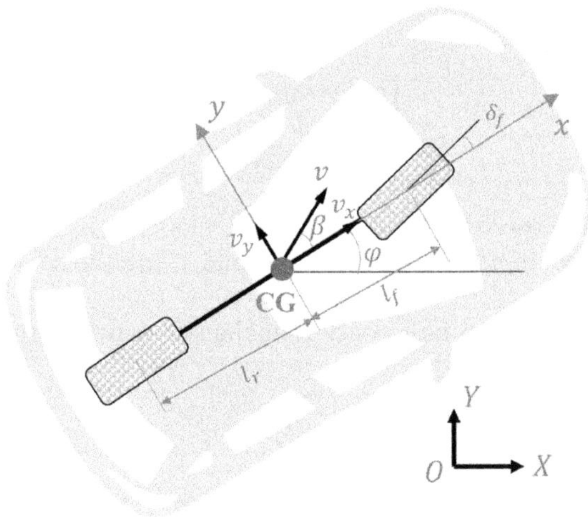

FIGURE 3.2: Bicycle kinematic model.

$$\Phi = \varphi + \beta \tag{3.8}$$

where the state vector and control vector of AV are denoted by $x = [v_x, \varphi, X_g, Y_g]^T$ and $u = [a_x, \delta_f]^T$, respectively. v_x, φ and Φ are the longitudinal velocity, yaw angle and heading angle of AV, respectively. (X_g, Y_g) is the coordinate position of the center of gravity (CG). a_x and δ_f denote the longitudinal acceleration and the front-wheel steering angle of AV. β denotes the sideslip angle of AV. l_f and l_r are the front and rear wheel bases of AV.

3.2.2 Vehicle Dynamic Model

3.2.2.1 Nonlinear Vehicle Dynamic Model

For the path-tracking controller design, the 3 DoF vehicle dynamic model is adopted, which includes lateral, yaw and roll motions. The diagram of the 3 DoF vehicle model is shown in Figure 3.3.

Based on Figure 3.3, the 3 DoF vehicle model is derived as follows.

$$mv_x(\dot{\beta} + r) + m_s h_s \ddot{\phi} = \sum F_y \tag{3.9}$$

$$I_z \dot{r} - I_{xz} \ddot{\phi} = \sum M_z \tag{3.10}$$

$$I_x \ddot{\phi} - I_{xz} \dot{r} = \sum L_x \tag{3.11}$$

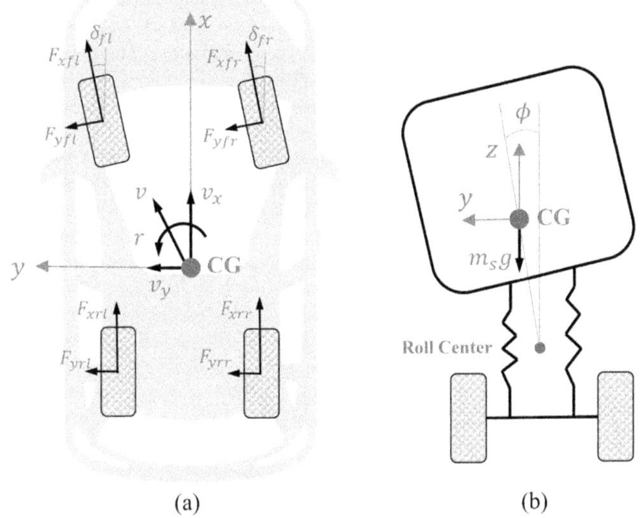

FIGURE 3.3: Nonlinear vehicle dynamic model: (a) top view; (b) rear view.

where r and ϕ denote the yaw rate and roll angle. m, m_s, and h_s denote the vehicle mass, vehicle sprung mass, and height of sprung mass. I_z, I_{xz}, and I_x denote the yaw inertia moment, the product of inertia, and the roll inertia moment, respectively. Moreover, $\sum F_y$, $\sum M_z$ and $\sum L_x$ denote the total lateral tire force, yaw moment and roll moment acting on the vehicle, which can be expressed as follows.

$$\sum F_y = F_{xfl}\sin\delta_{fl} + F_{xfr}\sin\delta_{fr} + F_{yfl}\cos\delta_{fl} + F_{yfr}\cos\delta_{fr} + F_{yrl} + F_{yrr} \tag{3.12}$$

$$\begin{aligned}\sum M_z = &[F_{xfl}\sin\delta_{fl} + F_{xfr}\sin\delta_{fr} + F_{yfl}\cos\delta_{fl} + F_{yfr}\cos\delta_{fr}]l_f \\ &-[F_{yrl} + F_{yrr}]l_r + [F_{yfl}\sin\delta_{fl} - F_{yfr}\sin\delta_{fr}]B/2\end{aligned} \tag{3.13}$$

$$\sum L_x = m_s g h_s \phi - b_\phi \dot{\phi} - k_\phi \phi \tag{3.14}$$

where $\delta_i(i = fl, fr, rl, rr)$ denotes the steering angle of each wheel, F_{xi} and $F_{yi}(i = fl, fr, rl, rr)$ denote the longitudinal force and lateral force of each tire, respectively. Besides, b_ϕ and k_ϕ denote the roll damping of vehicle suspension and the roll stiffness of vehicle suspension. B denotes the vehicle track.

Different tire models have been proposed to describe the tire-road contact force. Considering that Dugoff tire model can describe the tire-road contact force accurately with fewer parameters, it is applied in this study, which can be expressed as

$$F_{xi} = \frac{C_i s_i f(\lambda)}{1 - s_i}, \qquad F_{yi} = \frac{k_i \tan\alpha_i f(\lambda)}{1 - s_i} \tag{3.15}$$

$$s_i = \begin{cases} 1 - \frac{u_i}{R_w \omega_i}, & u_i < R_w \omega_i, \quad \omega_i \neq 0 \\ \frac{R_w \omega_i}{u_i} - 1, & u_i > R_w \omega_i, \quad u_i \neq 0 \end{cases} \tag{3.16}$$

$$f(\lambda) = \begin{cases} \lambda(2 - \mu) & \lambda < 1 \\ 1 & \lambda \geq 1 \end{cases} \tag{3.17}$$

$$\lambda = \frac{\mu F_{zi}(1 - s_i)}{2\sqrt{C^2 s_i^2 + k_i^2 \tan^2 \alpha_i}} \tag{3.18}$$

where s_i and $\alpha_i (i = fl, fr, rl, rr)$ denote the tire slip ratio and the tire slip angle, F_{zi} denotes the vertical tire force, μ is the tire-road friction coefficient, R_w is the effective rolling radius of the wheel, ω_i is the wheel spinning angular velocity and u_i is the longitudinal velocity at each wheel's center.

Additionally, the motion model for path tracking is built as follows.

$$\dot{\varphi} = r \tag{3.19}$$

$$\dot{X} = v_x \cos\varphi - v_y \sin\varphi \tag{3.20}$$

$$\dot{Y} = v_x \sin\varphi + v_y \cos\varphi \tag{3.21}$$

3.2.2.2 Linear Single-Track Model

To reduce the model complexity in motion prediction, the four-wheel vehicle model is simplified into a bicycle model. As Figure 3.4 shows, with the assumption of a small steering angle δ_f at the front wheel, it yields that $\sin\delta_f \approx 0$. Then the single-track bicycle model built for motion prediction can be expressed as follows [73, 221, 184].

$$\dot{x}(t) = \Gamma[x(t), u(t)] \tag{3.22}$$

$$\Gamma[x(t), u(t)] = \begin{bmatrix} v_y r + F_{xf} \cos\delta_f / m + F_{xr}/m \\ -v_x r + F_{yf} \cos\delta_f / m + F_{yr}/m \\ l_f F_{yf} \cos\delta_f / I_z - l_r F_{yr} / I_z \\ r \\ v_x \cos\varphi - v_y \sin\varphi \\ v_x \sin\varphi + v_y \cos\varphi \end{bmatrix} \tag{3.23}$$

where the state vector $x = [v_x, v_y, r, \varphi, X, Y]^T$, and the control vector $u = [a_x, \delta_f]^T$. v_x and v_y are the longitudinal and lateral velocities, respectively. r and φ are the yaw rate and yaw angle, respectively. (X, Y) is the coordinate position of the vehicle. $F_{xi}(i = f, r)$ and $F_{yi}(i = f, r)$ are the longitudinal and lateral tire forces of the front and rear wheels. l_f and l_r are the distances from the center of mass to the front axle and the rear axle, respectively. m is the vehicle mass, and I_z is the yaw moment of inertia.

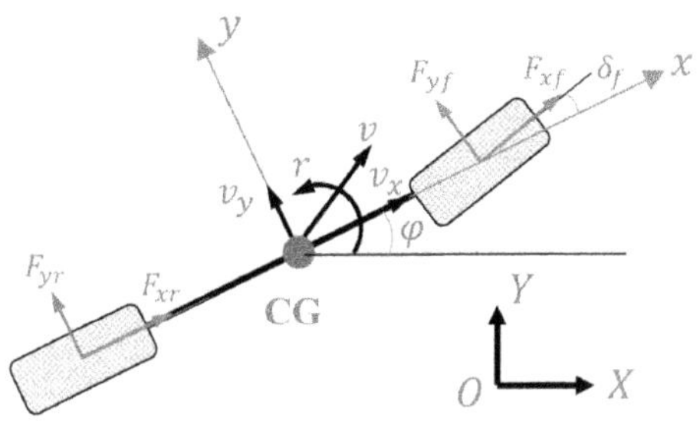

FIGURE 3.4: Linear single-track model.

Neglecting the air resistance and the rolling resistance, the longitudinal dynamics can be further simplified as

$$a_x = F_{xf} \cos \delta_f / m + F_{xr} / m \tag{3.24}$$

With the assumption of a small tire slip angle, the linear relationship between the lateral tire force and the tire slip angle can be achieved.

$$F_{yf} = -C_f \alpha_f, \quad F_{yr} = -C_r \alpha_r \tag{3.25}$$

where C_f and C_r are the cornering stiffness of the front and rear tires, respectively. Additionally, the slip angles of the front and rear tires α_f and α_r are given by

$$\alpha_f = -\delta_f + (v_y + l_f r)/v_x, \quad \alpha_r = (v_y - l_r r)/v_x \tag{3.26}$$

3.3 Driver Model

To realize human-like driving and decision making, a single-point preview driver model is proposed. As Figure 3.5 shows, point E is the current position of the driver. M is the predicted point along the current moving direction of the vehicle in the future, which is predicted by driver's brain according to the vehicle states and position. P is the preview point created by the driver's eyes, and consecutive preview points make up the planned path. The driver aims to minimize the distance between the predicted point and preview point. Finally, the purpose is realized by controlling the steering angle of the front wheel. The aforementioned content describes the basic working principle of driver model.

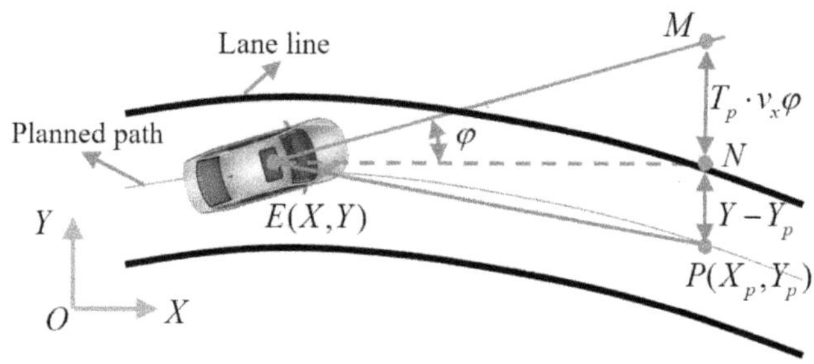

FIGURE 3.5: Driver model.

Considering the drivers driving characteristics, the driver model is expressed as follows [279].

$$\ddot{\delta}_f = -\frac{1}{aT_d}\dot{\delta}_f - \frac{1}{aT_d^2}\delta_f + \frac{K_s G_s}{aT_d^2}[Y_p - (Y + T_p \cdot v_x \varphi)] \tag{3.27}$$

where a is related to the damping rate of the model, T_d and T_p are the driver's physical delay time and predicted time, Y and Y_p are the lateral coordinate position of the vehicle and preview point, K_s is the transmission ratio of steering system, and G_s is the steering proportional gain.

The driver's driving style or characteristic is reflected by T_d, T_p and G_s. For instance, conservative drivers need speed more time for mental signal processing and muscular activation. Therefore, conservative drivers have larger T_d. However, aggressive drivers have reverse reactions [241].

Referring to [224, 291, 197], the parameters to reflect different driving styles are selected in Table 3.1.

TABLE 3.1: Parameters of Different Driving Styles

Parameters	Aggressive	Normal	Conservative
T_d	0.14	0.18	0.24
T_p	1.02	0.94	0.83
G_s	0.84	0.75	0.62
a	0.24	0.23	0.22

To analyze the characteristic of the driver model with different driving styles, the distance error between predicted point M and preview point P is denoted by ΔY, i.e., $\Delta Y = Y_p - (Y + T_p \cdot v_x \varphi)$. Then, the transfer function from ΔY to δ_f is derived as

$$\delta_f(s) = \frac{K_s G_s}{a T_d^2 s^2 + T_d s + 1} \Delta Y(s) \tag{3.28}$$

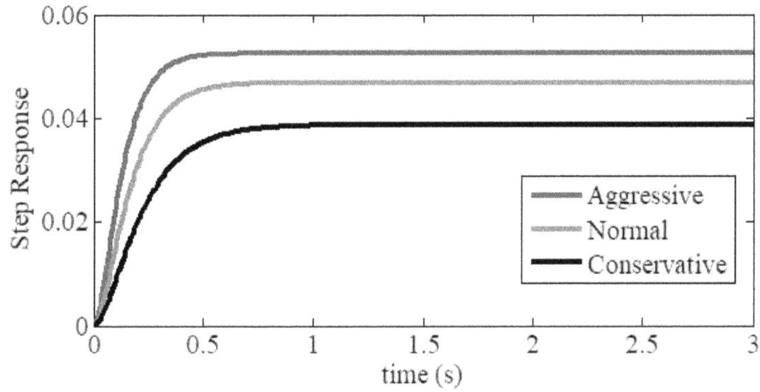

FIGURE 3.6: Step response results of the driver model under three driving styles.

According to (3.28), Figure 3.6 presents the step response results of the driver model under three different driving styles. It can be seen that aggressive style has fastest response speed and maximum stable gain. By comparison, conservative style has longest response time and minimum stable gain. The response result of normal style is between the above two driving styles. From the aforementioned analysis, it can be concluded that aggressive drivers usually control the steering wheel with large amplitude and high frequency. However, conservative drivers show the opposite behaviors.

3.4 Integrated Model for Human-Like Driving

Combining the driver model and the vehicle-road model yields the integrated model for human-like driving. As Figure 3.7 shows, the integrated model has two inputs, i.e., a_x and Y_p. a_x is input of vehicle-road model and Y_p is input of driver model, which wait for solution. The outputs of the integrated model are vehicle states and position, which is used for decision making and motion planning.

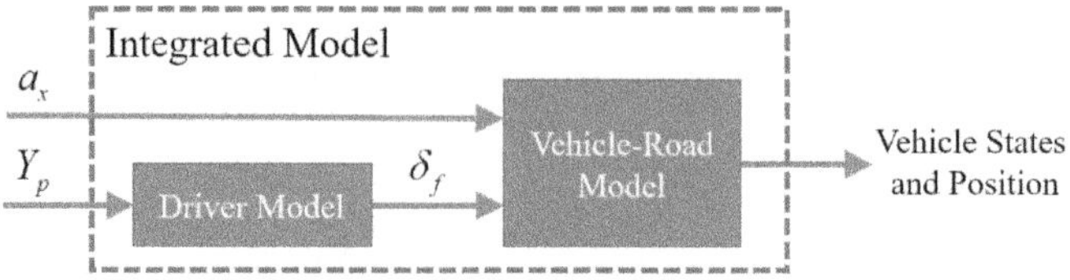

FIGURE 3.7: Integrated model.

According to the linear single-track model and driver model, the integrated model for human-like driving is expressed as.

$$\dot{x}(t) = f[x(t), u(t)] \tag{3.29}$$

$$f[x(t), u(t)] = \begin{bmatrix} v_y r + a_x \\ -v_x r + F_{yf}\cos\delta_f/m + F_{yr}/m \\ l_f F_{yf}\cos\delta_f/I_z - l_r F_{yr}/I_z \\ r \\ v_x\cos\varphi - v_y\sin\varphi \\ v_x\sin\varphi + v_y\cos\varphi \\ \dot{\delta}_f \\ -\frac{1}{aT_d}\dot{\delta}_f - \frac{1}{aT_d^2}\delta_f + \frac{K_s G_s}{aT_d^2}[Y_p - (Y + T_p \cdot v_x\varphi)] \end{bmatrix} \tag{3.30}$$

where the state vector $x = [v_x, v_y, r, \varphi, X, Y, \delta_f, \dot{\delta}_f]^T$, and the control vector $u = Y_p$.

3.5 Conclusion

In this chapter, system modeling is conducted for decision making, motion prediction, planning and control. Two kinds of vehicle model, i.e., vehicle kinematic model and vehicle dynamic model, are proposed. In the construction of vehicle kinematic model, the simplified mass point kinematic model is designed for the path planning of AVs, and the simplified bicycle kinematic model is designed for the motion prediction and decision making of AVs. Besides, the nonlinear and linear vehicle dynamic models are used for the motion controller design of AVs. Furthermore, the vehicle dynamic model is combined with the driver model, yielding an integrated model for human-like driving. With different parametric settings in the driver model, the integrated model is in favor of realizing different driving styles and personalized decision making.

Chapter 4

Motion Planning and Tracking Control of Autonomous Vehicles

DOI: 10.1201/9781003287087-4

4.1 Background

The motion planning of AVs can be divided into the longitudinal planning and the lateral planning. In detail, the longitudinal planning is the planning of longitudinal velocity or acceleration, which is considered in the decision-making module in this book. The lateral planning is the trajectory planning or path planning. The difference is the planned path is only related to the position coordinates. However, the planned trajectory is associated with time. In general, trajectory planning is usually used to deal with the local planning, especially realizing collision avoidance of moving obstacles. Path planning is usually applied to the global planning, especially realizing collision avoidance of static obstacles. Although all kinds of algorithms have been applied to the motion planning of AVs, human-like driving characteristics are seldom considered. Human-like motion planning for AVs is in favor of decreasing the misunderstanding from human drivers. In this chapter, the social behaviors of surrounding traffic occupants are considered in the motion planning of AVs, and the artificial potential field (APF) approach is combined with MPC to realize human-like trajectory planning. Moreover, to deal with the path planning of AVs on unstructured roads, the visibility graph method is combined with NMPC to realize collision avoidance planning of both static and moving obstacles.

As to the motion control issue of AVs, path tracking control is the main control mission. As introduced in Chapter 1, different control algorithms have been applied to the path tracking control of AVs. Most control algorithms can realize good path tracking in normal conditions. With the increase of vehicle speed, the active safety issue becomes more and more important. In addition to the path tracking issue, the issues regarding handling stability and rollover prevention are very critical. To find a good balance between the path tracking performance and other active safety performances becomes a challenge. The multi-objective coordinated control is an effective solution to this issue.

Moreover, the strong robustness is required for the designed controller to deal with extreme driving conditions, e.g., low road friction coefficient condition and large curvature path tracking condition. Considering the advantages in the field of multi-objective and multi-constraint optimization, the LTV MPC approach is applied to the motion controller design of AVs in this chapter.

4.2 Human-Like Trajectory Planning for AVs on Highways

In this section, the social behaviors of surrounding traffic occupants are considered in the motion planning of AVs on highways. Reflected by driving styles and intentions of surrounding vehicles, the social behaviors are taken into consideration during the modeling process. The APF method is adopted in the motion planning model, which uses different potential functions to describe surrounding vehicles with different behaviors and road constraints. Finally, MPC is utilized for motion prediction and to solve the motion planning issue of AVs.

4.2.1 Artificial Potential Field Model

4.2.1.1 APF Model for Vehicles

APF approach has the advantage to describe the size and risk of the obstacle vehicles with the APF value and distribution. The optimal planned path or trajectory usually has the smallest APF value. In this section, the APF approach is combined with the MPC optimization for trajectory planning of AVs.

For the obstacle vehicle, the APF distribution $P^{ov}(X,Y)$ at the position (X,Y) can be described with the following equations [215].

$$P^{ov}(X,Y) = a^{ov}e^{\eta} \tag{4.1}$$

$$\eta = -[\frac{\hat{X}^2}{2\rho_X^2} + \frac{\hat{Y}^2}{2\rho_Y^2}]^b + cv_x ov\xi \tag{4.2}$$

$$\xi = k^{ov}\frac{\frac{\hat{X}^2}{2\rho_X^2}}{\sqrt{\frac{\hat{X}^2}{2\rho_X^2} + \frac{\hat{Y}^2}{2\rho_Y^2}}} \tag{4.3}$$

$$k^{ov} = \begin{cases} -1, & \hat{X} < 0 \\ 1, & \hat{X} \geq 0 \end{cases} \tag{4.4}$$

$$\begin{bmatrix} \hat{X} \\ \hat{Y} \end{bmatrix} = \begin{bmatrix} \cos\varphi^{ov} & \sin\varphi^{ov} \\ -\sin\varphi^{ov} & \cos\varphi^{ov} \end{bmatrix} \begin{bmatrix} X - X^{ov} \\ Y - Y^{ov} \end{bmatrix} \tag{4.5}$$

where (X^{ov}, Y^{ov}) is the CG position of the obstacle vehicle. φ^{ov} and v_x^{ov} denote the heading angle and longitudinal velocity. ρ_X and ρ_Y are the convergence coefficients along the directions of X and Y, respectively. b denotes the shape coefficient.

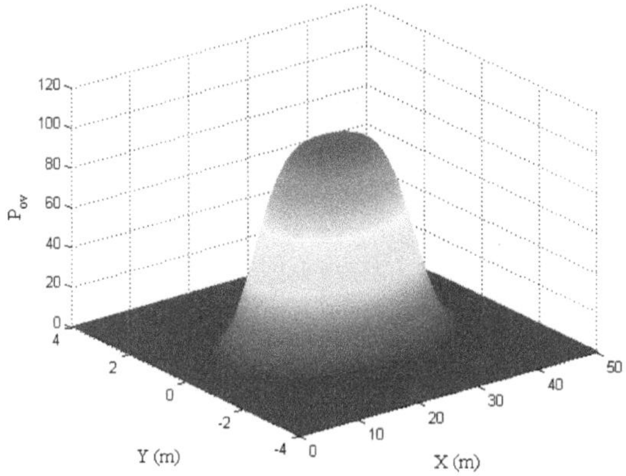

FIGURE 4.1: 3D map of the APF model for the obstacle vehicle.

Based on above equations, the APF distribution of the obstacle vehicle is displayed in Figure 4.1. Red color means large APF value, which is more dangerous for trajectory planning. Blue color means low APF value, which is safer for trajectory planning.

4.2.1.2 APF Model for Road

Besides the vehicle, the APF model can be applied to the road, which is expressed as follows.

$$P^r(X,Y) = a^r e^{-d+d^r+0.5W} \tag{4.6}$$

where a^r denotes the maximum value for the APF distribution of the road, d denotes the minimum distance from the position (X,Y) to the lane mark, d^r denotes the safety coefficient, and W denotes the vehicle width.

Based on the above equation, the APF distribution for the three-lane road is displayed in Figure 4.2.

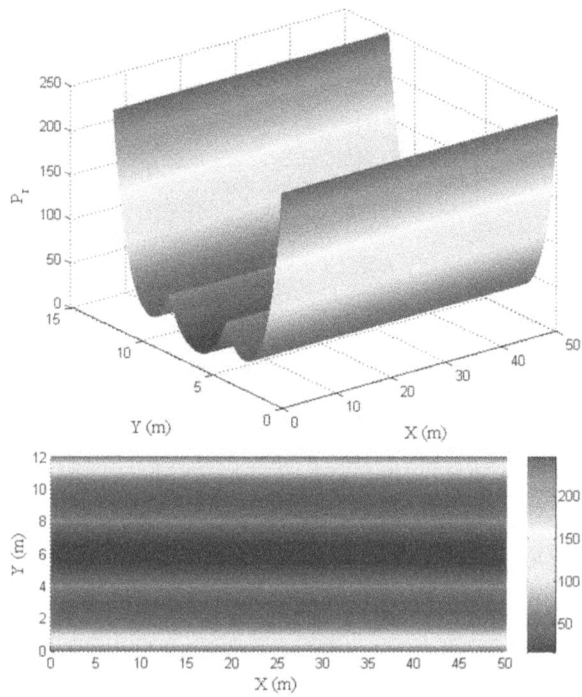

FIGURE 4.2: 3D map of the APF model for road with three lanes.

4.2.1.3 Integrated APF Model for Trajectory Planning

The combination of the APF models of all obstacle vehicles and roads yields the integrated APF model, which is derived as follows.

$$P^c(X,Y) = \sum_{i=1}^{m} P_i^{ov}(X,Y) + \sum_{j=1}^{n} P_j^r(X,Y) \tag{4.7}$$

where m and n denote the number of all obstacle vehicles and the number of lane lines, respectively.

Taking the three vehicles on a three-lane highway as an instance, the integrated APF model is illustrated in Figure 4.3. It can be found that the free space has lower APF value, which indicates these areas are safer for trajectory planning.

4.2.2 Trajectory Planning Considering Different Social Behaviors of Obstacle Vehicles

As mentioned in Chapter 2, different obstacle vehicles have various driving obstacles. To make the trajectory planning algorithm adaptive to different

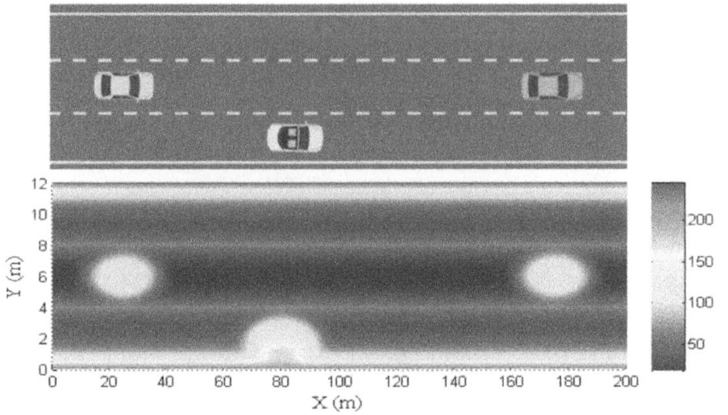

FIGURE 4.3: APF model of three vehicles on a highway road with three lanes.

driving styles, we give the adaptive APF model for trajectory planning, which is shown in Figure 4.4. It can be found that three different driving styles have various APF distributions. The aggressive driving style has the widest APF distribution along the moving direction, which means that we should consider larger safe margin for the collision avoidance of the aggressive vehicle. However, the conservative driving style has the smallest APF distribution along the moving direction, indicating that the trajectory planning is safer when facing the conservative obstacle vehicle.

4.2.3 Trajectory Planning with APF Considering Trajectory Prediction

The mass point kinematic model (3.2) is used for trajectory planning. For the host vehicle (HV), the state vector $x = [v_x^{hv}, \varphi^{hv}, X^{hv}, Y^{hv}]^T$ and the control vector $u = a_y^{hv}$.

For trajectory planning, (3.2) rewritten as a discretized system.

$$x(k+1) = x(k) + F[x(k), u(k)] \tag{4.8}$$

Additionally, the output vector y is expressed with integrated APF model.

$$y(k) = g[x(k), u(k)] = P^c(X^{hv}(k), Y^{hv}(k)) \tag{4.9}$$

Furthermore, the MPC approach is applied to the output prediction. The

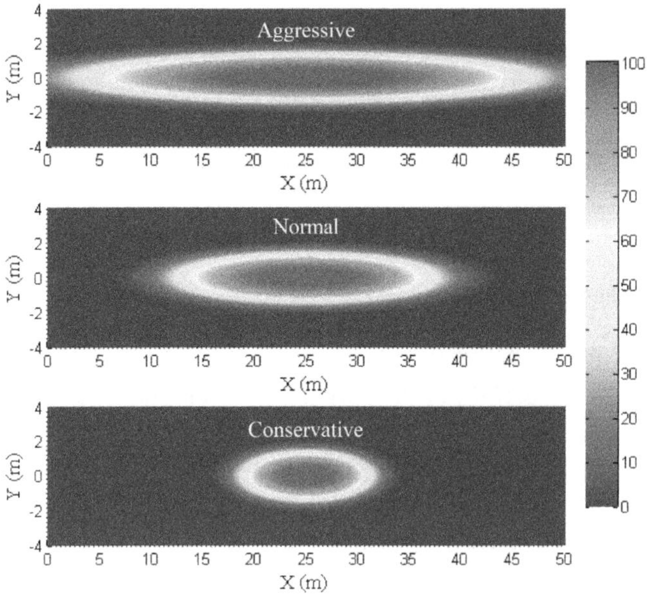

FIGURE 4.4: APF model considering with different driving styles.

predicted outputs at the time step k are derived as follows.

$$\begin{aligned}
y(k+1|k) &= g[x(k+1|k), u(k|k)] \\
y(k+2|k) &= g[x(k+2|k), u(k+1|k)] \\
&\vdots \\
y(k+N_c|k) &= g[x(k+N_c|k), u(k+N_c-1|k)] \\
y(k+N_c+1|k) &= g[x(k+N_c+1|k), u(k+N_c-1|k)] \\
&\vdots \\
y(k+N_p|k) &= g[x(k+N_p|k), u(k+N_c-1|k)]
\end{aligned} \tag{4.10}$$

Defining the output sequence and control sequence as follows.

$$\mathbf{y}(k) = [y(k+1|k), y(k+2|k), \cdots, y(k+N_p|k)]^T \tag{4.11}$$

$$\mathbf{u}(k) = [u(k|k), u(k+1|k), \cdots, u(k+N_c-1|k)]^T \tag{4.12}$$

where N_p and N_c denote the prediction horizon and control horizon, $N_p > N_c$.

Considering the control consumption and performance indexes, the cost function of trajectory planning is constructed as follows.

$$J^p(k) = \mathbf{y}^T(k)Q_1\mathbf{y}(k) + \Delta\mathbf{Y}^T(k)Q_2\Delta\mathbf{Y}(k) + \mathbf{u}^T(k)R\mathbf{u}(k) \tag{4.13}$$

where Q_1 and Q_2 denote the performance weighting matrices, R denotes the control consumption weighting matrix, $\Delta\mathbf{Y}$ denotes the lateral distance error sequence between the planned trajectory and the lane center.

Then, the trajectory planning issue is converted into an optimization problem with multiple constraints.

$$\mathbf{u}(k) = \arg\min_{\mathbf{u}(k)} J^p(k) \tag{4.14}$$

subject to $u_{\min} \leq u(k+i-1) \leq u_{\max}$, where $u_{\min}$ and $u_{\max}$ are the low and high control boundaries.

Finally, the optimal control sequence can be worked out.

$$\mathbf{u}^*(k) = [u^*(k|k), u^*(k+1|k), \cdots, ^* u(k+N_c-1|k)]^T \tag{4.15}$$

The first vector $u^*(k|k)$ of the optimal control sequence is used for trajectory planning. At the next time step $k+1$, a new optimization is started over a shifted prediction horizon again with the updated state $x(k+1|k+1)$.

4.2.4 Simulation and Discussion

To verify the APF-based human-like trajectory planning algorithm, three test cases are designed and carried out considering different driving styles of surrounding obstacle vehicles. The detailed setting is illustrated in Figure 4.5. In favor of the trajectory planning verification, the Stackelberg game theoretic approach is used for decision making, which is studied in Chapter 5. It can be found from Figure 4.5 that the complexity of the test scenario increases from Case 1 to Case 3, in favor of evaluating the generalizability of the trajectory planning algorithm in different scenarios. The algorithm verification is conducted with the Matlab-Simulink platform. The initial velocities and positions of all vehicles are presented in Figure 4.5.

4.2.4.1 Testing Case 1

In Case 1, a single lane-change scenario is designed on a two-lane highway. HV and vehicle 1 (V1) move on the right lane and vehicle 2 (V2) moves on the left lane. For HV, V1 is the lead vehicle and V2 is the obstacle vehicle. Due to the low-speed driving behavior of V1, HV must make the decision, slowing down and following the lead vehicle, or changing lanes and overtaking. If conducting lane-changing, HV must interact and game with V2. The driving behavior and driving style of V2 have a significant effect on the decision-making result of HV. In this case, V2 is defined with three kinds of driving styles, i.e., aggressive, normal and conservative. To simplify the test scenario, V1 is assumed to move with an constant speed. The test results are displayed from Figures 4.6 to 4.8.

From the test results, we can find that different driving styles of V2 will lead to different decision making and planning results of HV. If V2 is aggressive, V2 would fight for the right of way and not be willing to give way for HV.

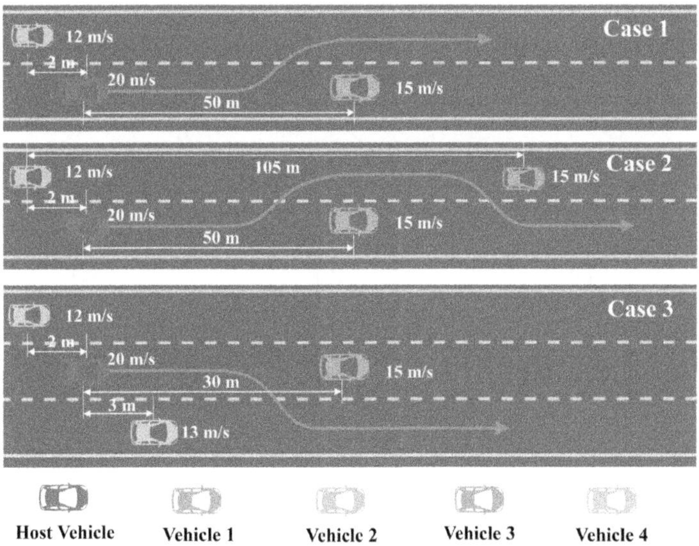

FIGURE 4.5: Three test cases for decision making and motion planning.

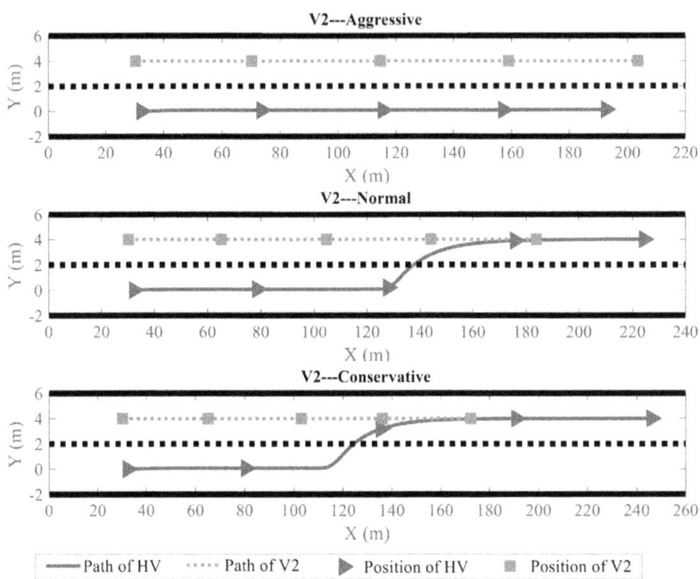

FIGURE 4.6: Results of the decision making and trajectory planning considering different driving styles of obstacle vehicles in Case 1.

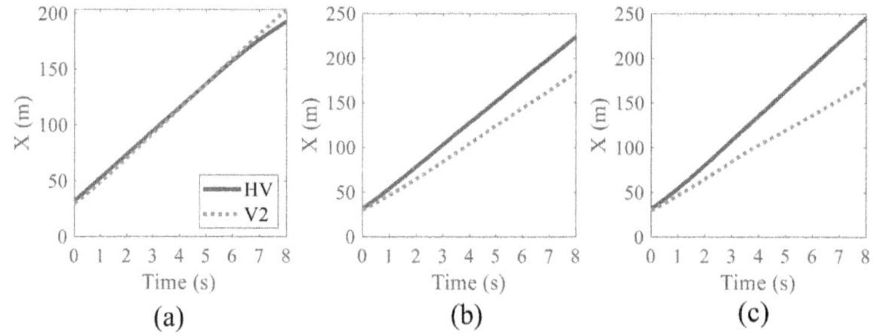

FIGURE 4.7: Testing results of vehicle longitudinal positions in Case 1: (a) V2 is aggressive; (b) V2 is normal; (c) V2 is cautious.

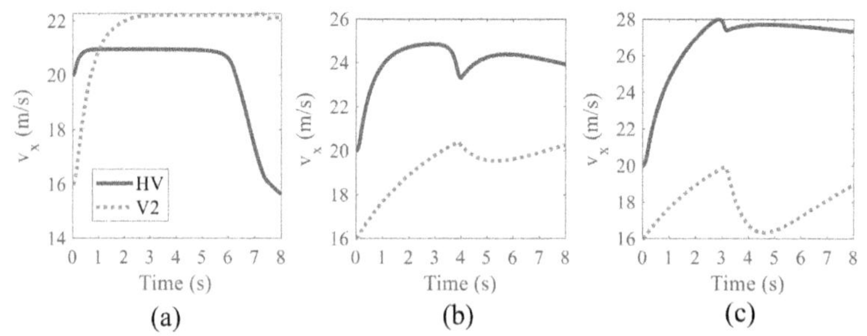

FIGURE 4.8: Testing results of vehicle velocities in Case 1: (a) V2 is aggressive; (b) V2 is normal; (c) V2 is cautious.

As a result, V2 chose a sudden accelerating behavior. HV gave up overtaking and had to slow down and keep the safe distance between V1. From Figure 4.8, we can find an obvious deceleration of HV after the sixth second. However, if V2 is the normal driving style, it would lead to a different decision making and trajectory planning result of HV. HV finished the accelerating and overtaking behavior perfectly. V1 made the decision decelerating and gave way for HV after the fourth second. If the driving style of V1 is conservative, the similar test results are obtained. V1 slowed down and gave way for HV. HV speeded up and finished the overtaking mission. The difference is that the lane-change starting time is earlier compared to the second scenario. V1's decelerating behavior is more obvious. The gap between V1 and HV is larger, which indicates that the lane-change process of HV is much safer.

4.2.4.2 Testing Case 2

This case simulate a double lane-change scenario on two-lane highways, which is an improvement of Case 1. In this case, HV and V1 are moving on the right lane, and V2 and V3 are moving on the left lane. We assume that V2 is normal, and HV has finished the overtaking behavior. After lane-change, V3 becomes the lead vehicle of HV. However, due to the low-speed driving of V3, HV must make the decision slowing down and following V3, or accelerating and conduct the lane-change again. If overtaking again, V1 will be the obstacle of HV. HV must interact with V1. V1's driving behavior will have significant effect on HV's decision making and trajectory-planning results. In this case, V3 is assumed to move forward with a constant velocity of 15 m/s. V1 has three different driving styles, i.e., aggressive, normal, and conservative. The test results are presented in Figures 4.9 to 4.11.

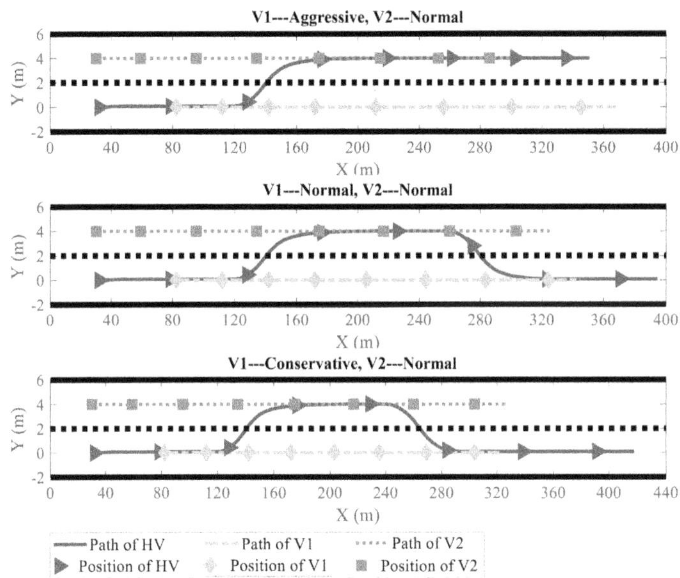

FIGURE 4.9: Results of decision making and trajectory planning considering different driving styles of obstacle vehicles in Case 2.

From the test results, we can conclude that different driving styles of obstacle vehicles result in different test results regarding decision making and trajectory planning. It can be found from Figure 4.11 that although V1 moves forward at a constant speed at first, it will have reactive behaviors if it finds HV has a lane-change intention. If V1 is aggressive, it will speed up and fight for the way. As a result, HV has to keep the lane and slow down to keep a safe gap between itself and V3. If V1's driving style is normal, it will not fight for the right of way against HV. Once HV moves ahead and has a larger speed, the safe overtaking will be finished easily. If V1's driving style is conservative, it will be safer and easier for HV to finish the double lane change behavior. We can see from Figure 4.9 that HV has smaller double lane change distance due to the conservative driving style of V1.

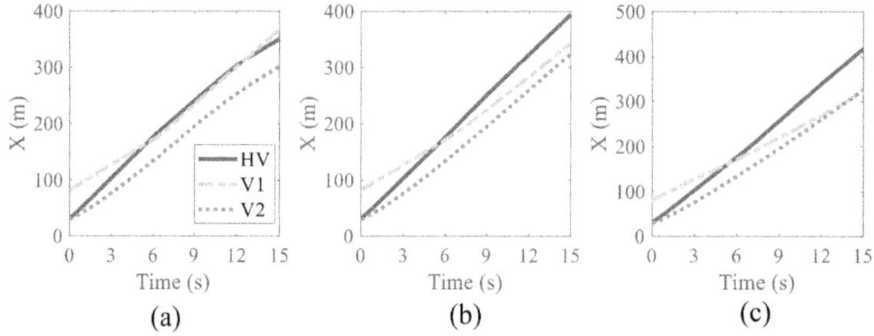

FIGURE 4.10: Testing results of vehicle positions in Case 2: (a) V1 is aggressive and V2 is normal; (b) V1 is normal and V2 is normal; (c) V1 is cautious and V2 is normal.

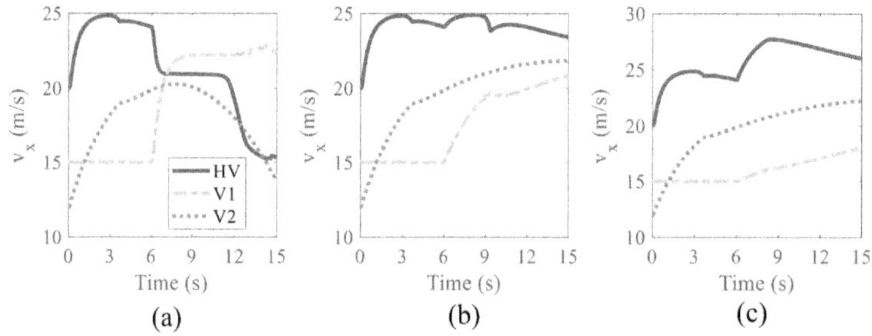

FIGURE 4.11: Testing results of velocities in Case 2: (a) V1 is aggressive and V2 is normal; (b) V1 is normal and V2 is normal; (c) V1 is cautious and V2 is normal.

4.2.4.3 Testing Case 3

Case 3 presents a three-lane highway scenario, which is more complex that the above two cases. In this case, V2 moves on the left lane, HV and V1 move on the middle lane, and V4 moves on the right lane. For HV, V1 is the lead vehicle, which is assumed to move forward with a constant velocity of 15 m/s. V2 and V4 are obstacle vehicles. Due to the low-speed driving behavior of V1, HV has to make decisions, i.e., slowing downing and following V1, or accelerating and changing lanes. If changing lanes, HV must make the decisions that change to the left or the right. It must consider the driving behaviors of V2 and V4. Namely, the driving styles of V2 and V4 have significant effects on the decision making and trajectory-planning results of HV. To this end, different driving styles are set for V2 and V4 in this case. The test results are illustrated in Figures 4.12 to 4.14.

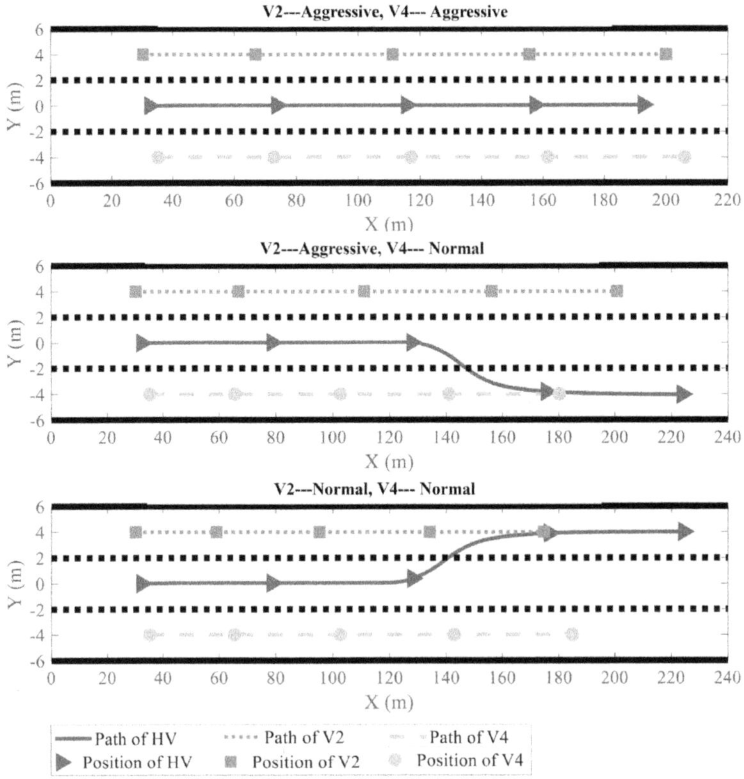

FIGURE 4.12: Results of decision making and trajectory planning considering different driving styles of obstacle vehicles in Case 3.

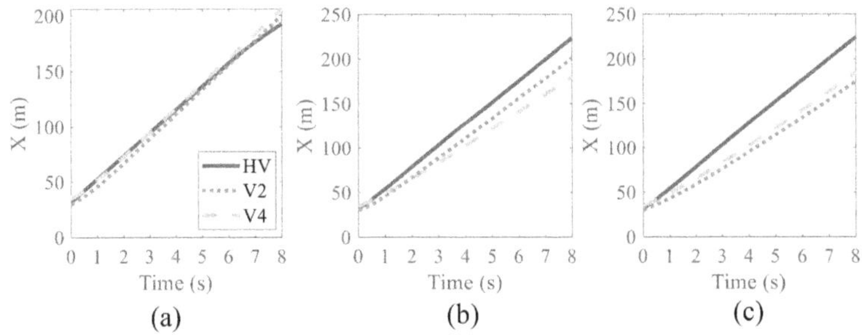

FIGURE 4.13: Testing results of vehicle positions in Case 3: (a) V2 is aggressive and V4 is aggressive; (b) V2 is aggressive and V4 is normal; (c) V2 is normal and V4 is normal.

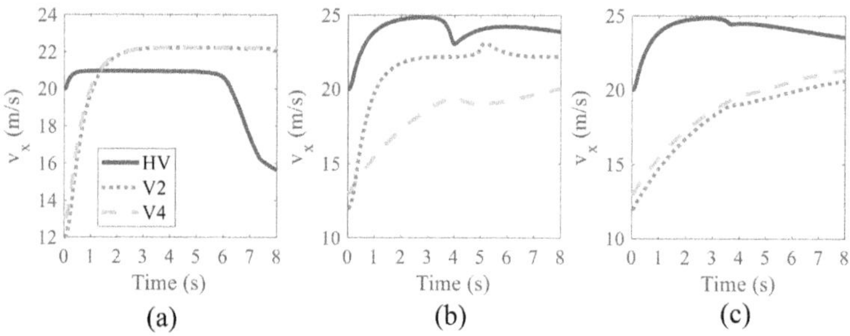

FIGURE 4.14: Testing results of vehicle velocities in Case 3: (a) V2 is aggressive and V4 is aggressive; (b) V2 is aggressive and V4 is normal; (c) V2 is normal and V4 is normal.

We can find that different driving styles of V2 and V4 cause different test results if HV. In the first scenario, both V2 and V4 are aggressive. As a result, both V2 and V4 speed up and are unwilling to give ways for HV. HV has to choose lane-keeping and slowing down. After the sixth second, HV has a sudden decelerating action to keep the safe distance between itself and the lead vehicle. In the second scenario, V2 is still aggressive and unwilling to give ways for HV. However, due to the normal driving style of V4, V4 chooses to give ways for HV, and HV successfully changes the lane to the right. In the third scenario, both V2 and V4 and normal. HV would choose the left-side lane change due to the smaller cost generated on the left lane. It can be explained from Figure 4.14 that the velocity of V2 is smaller than that of V4.

4.2.5 Summary

This section presents a human-like trajectory planning approach for AVs. Social behaviors, which are reflected by three different driving styles of obstacle vehicles, are defined. The APF model is used to describe the driving characteristics of surrounding obstacle vehicles and applied to the trajectory planning of AVs. Besides, the MPC optimization is used for the APF value prediction. Then, the trajectory planning issue is transformed into a closed-loop interactive optimization problem with multiple constraints. Finally, three test cases are designed and carried out to verify the human-like trajectory planning algorithm. Testing results indicate that the proposed algorithm is able to deal with reasonable and safe trajectory planning for AVs under various social behaviors of surrounding obstacle vehicles.

4.3 Path Planning of AVs on Unstructured Roads

This section presents an integrated path planning algorithm for AVs on unstructured roads. The visibility graph method is effective to conduct global path planning for AVs, which is used for collision avoidance of static obstacles. However, the planned path is not acceptable for path tracking control due to the large curvature at the turning position. Besides, the visibility graph method cannot deal with moving obstacles. To address the above issue, NMPC is utilized to conduct the second planning considering various constraints, e.g., minimum turning radius, safe distance, control constraint, tracking error, etc. Moreover, the polynomial fitting approach is used to predict the moving trajectories of moving obstacles. Finally, three simulation cases are conducted to evaluate the performance of the proposed planning algorithm, verifying the feasibility and effectiveness in different driving conditions.

4.3.1 Problem Statement

If point S and point F denote the start point and the final point of AV, the path planning issue can be described as finding the shortest safe path from point S to point F. Safety and efficiency are two critical performance indexes for the path planning algorithm. In the path planning process, AVs usually face two kinds of obstacles, i.e., static obstacles and moving obstacles. Due to the fixed position, the collision avoidance of static obstacles is very easy to realize with existing planning algorithms. Compared with the collision avoidance of static obstacles, the collision avoidance of moving obstacles is more difficult considering the motion uncertainties of moving obstacles. Due to the limitation of the single planning algorithm, the comprehensive planning algorithm that consists of multiple planning algorithm is usually used for the mixed collision avoidance of both static and moving obstacles. Moreover, there exists obvious difference of path planning in the low-speed driving condition and the high-speed driving condition. In the high-speed driving condition, the lateral stability issue of AVs must be taken into consideration. For instance, if the planned path has large curvature at the turning position, it will lead to large lateral acceleration, which is harmful to driving safety. As a result, the vehicle may lose stability. Additionally, the motion uncertainty of moving obstacles is another challenge for the dynamic planning of AVs.

To deal with the above issues, the visibility graph method is utilized to plan the global path in the driving environment of static obstacles. To further improve the quality of the planned path, all kinds of constraints, e.g., minimum turning radius, safe distance, and control constraint, are considered in the second planning process with NMPC. To realize dynamic planning regarding the moving obstacles, considering motion uncertainties, the polynomial fitting approach is used to predict the moving trajectories of moving obstacles. Then, the predicted trajectories are outputted to the path planner to improve the planning accuracy. Finally, the comprehensive path planning algorithm that integrates the visibility graph method and NMPC is capable of the static and dynamic path planning.

4.3.2 Path Planning for Static Obstacles

4.3.2.1 Pretreatment of Driving Scenario

Considering the constraints of body size and the minimum turning radius, AVs cannot go through or turn around in a narrow space. Therefore, it is necessary to pretreat the driving scenario before conducting the planning process, e.g., merging the obstacles together. As Figure 4.15 shows, the red car (AV) is preparing to enter the narrow gap between the obstacles O_1 and O_2. The issue that if the AV can go through the gap depends on many factors, including the body size, the minimum turning radius R_{min}, the width of the narrow gap W, the moving direction and the initial position of AV, etc. To simplify the issue, AV is simplified as a particle P, moving along the x axis. In Figure 4.15, D denotes the distance from P to the obstacles O_1. O_{11} and O_{21} denote the vertexes of the obstacles O_1 and O_2, respectively. The coordinates of points O_{11} and O_{21} are defined as $(0, D)$ and (W, D), respectively.

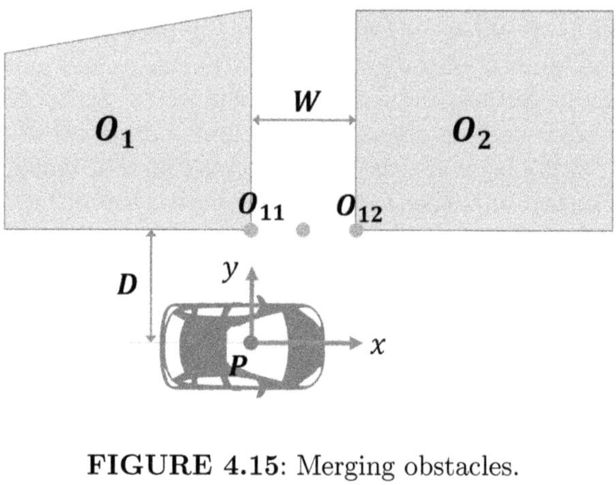

FIGURE 4.15: Merging obstacles.

Assuming that the AV does not change its moving direction before the turning point, the following conclusions are effective. (1) $D \geq R_{min}$, the AV can enter the narrow gap safely and the turning point is (x_t, y_t), where $x_t \in (-R_{min}, W - R_{min})$ and $y_t = 0$. (2) $D < R_{min}$ and $W > R_{min}$, the AV can enter the narrow gap safely and the turning point is (x_t, y_t), where $x_t \in (-\sqrt{R_{min}^2 - (R_{min} - D)^2}, W - R_{min})$ and $y_t = 0$. (3) $D < R_{min}$ and $R_{min} - \sqrt{R_{min}^2 - (R_{min} - D)^2} < W \leq R_{min}$, the AV can enter the narrow gap safely and the turning point is (x_t, y_t), where $x_t \in$

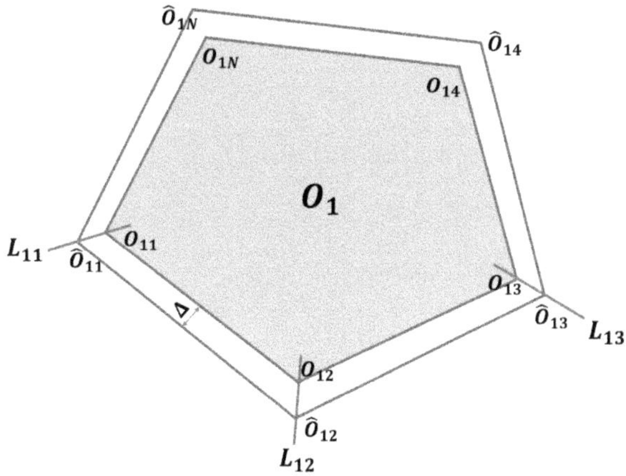

FIGURE 4.16: Extending boundary of obstacles.

$(-\sqrt{R_{min}^2 - (R_{min} - D)^2}, -(R_{min} - W))$ and $y_t = 0$. (4) $D < R_{min}$ and $W < R_{min} - \sqrt{R_{min}^2 - (R_{min} - D)^2}$, the AV cannot enter the narrow gap. As a result, the two obstacles should be merged as one obstacle before conducting path planning.

To further advance the safety in the path planning process, the safe distance between the AV and the obstacle is very necessary. To this end, the boundaries of obstacles are usually extended for path planning. As Figure 4.16 shows, the boundary of obstacle O_1 is extended with a safe distance Δ. Before extending the boundaries, all the vertices $O_{11}, O_{12}, \cdots, O_{1N}$, are detected. Then, the original boundaries of obstacle O_1, i.e., $O_{11}O_{12}, \cdots, O_{1N-1}O_{1N}$ can be worked out. Based on the extended distance Δ, the new boundaries can be obtained. Finally, the points of intersection of all new boundaries, i.e., $\hat{O}_{11}, \hat{O}_{12}, \cdots, \hat{O}_{1N}$, are the new vertices, which are used for path planning of AVs.

4.3.2.2 Path Planning with the Visibility Graph Method

After finishing the pretreatment of static obstacles, the visibility graph method is used to plan the global path. As Figure 4.17 shows, there exist four obstacles, i.e., O_1, O_2, O_3, and O_4. The AV wants to move from the start point S to the final point F. It must address the four obstacles. We can see that the boundaries of the four obstacles have been extended. All obstacles are transformed into polygonal obstacles with the extended boundaries and viewed as graphs. All the feasible paths between vertices have been drawn in the graph, which yields a feasible path network. The visibility graph method aims to find the shortest path in the graph from the start point S to the final point F. The visibility graph method can be described as $\langle \Xi_s, \Xi_{init}, \Xi_{goal}, \Xi_{obst} \rangle$, where Ξ_s is the given searching space, Ξ_{init} is the initial position, $\Xi_{init} \in$

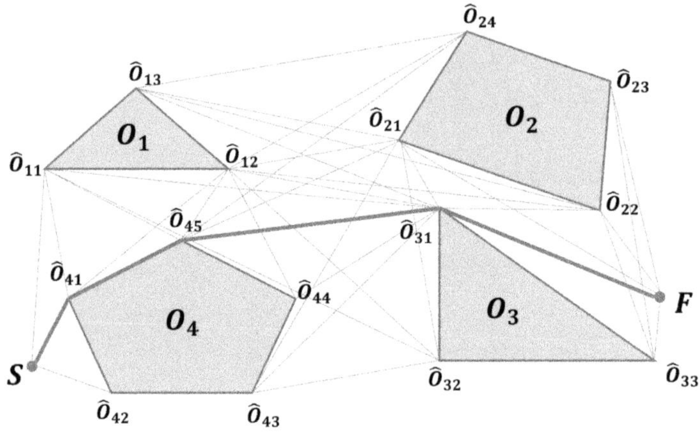

FIGURE 4.17: Visibility graph method.

Ξ_s, Ξ_{goal} is the goal position, $\Xi_{goal} \in \Xi_s$, and Ξ_{obst} is the set of obstacles, $\Xi_{obst} \in \Xi_s$. If the path sequence $\varsigma = [\varsigma_1, \varsigma_2, \cdots, \varsigma_k]$ is a feasible path in the graph, it yields that $\varsigma_1 = \Xi_{init}$, $\varsigma_k = \Xi_{goal}$ and $\varsigma \cap \Xi_{obst} = 0$. In Figure 4.17, $[S, \hat{O}_{41}, \hat{O}_{45}, \hat{O}_{31}, F]$ is the feasible path sequence for this issue, and the red path the shortest path using the visibility graph method.

4.3.2.3 Path Optimization Using NMPC

As Figure 4.17 shows, the planned path is made up of a set of line segments. Although the planned path with the visibility graph method is the shortest safe path, some issues are not considered, e.g., the smoothness of the path, and the minimum turning radius of AVs. If these issue are not addressed, the planned path is unacceptable for path tracking control. To this end, the quality of the planned path should be further improved. In this subsection, the second planning issue of the path planned by the visibility graph method is transformed into an optimization issue, and NMPC is used to conduct the nonlinear optimization considering multiple constraints.

To conduct the second planning, the mass point kinematic model (3.2) is utilized. (3.2) is rewritten in the discrete form.

$$\begin{cases} x(k+1) = x(k) + F[x(k), u(k)] \\ y(k) = g[x(k), u(k)] \end{cases} \tag{4.16}$$

The control boundaries of u is expressed as follows.

$$u_{min} \leq u \leq u_{max} \tag{4.17}$$

where u_{min} and u_{max} denote the minimum and maximum values of u.

Besides, the error boundaries are defined as follows.

$$\Delta y_{min} \leq y - y_{ref} \leq \Delta y_{max} \tag{4.18}$$

where y_{ref} denotes the reference path planned by visibility graph method, Δy_{min} and Δy_{max} denote the minimum and maximum tracking errors, which are selected according to the driving environment. For instance, the tracking error should be increased at the turning point.

Besides, the lateral stability of AV is reflected by the lateral acceleration, which is directly affected by the minimum turning radius. Therefore, it is necessary to take the minimum turning radius into consideration during the path optimization process. The path curvature ϱ is expressed as follows.

$$\varrho = \frac{|\dot{X}\ddot{Y} - \ddot{X}\dot{Y}|}{(\dot{X}^2 + \dot{Y}^2)^{3/2}} \tag{4.19}$$

To express the derivative of X and Y in the discrete form for NMPC optimization, it yields that

$$\dot{X} = (X(k+1) - X(k))/T \tag{4.20}$$

$$\dot{Y} = (Y(k+1) - Y(k))/T \tag{4.21}$$

$$\ddot{X} = (X(k+2) - 2X(k+1) + X(k))/T^2 \tag{4.22}$$

$$\ddot{Y} = (Y(k+2) - 2Y(k+1) + Y(k))/T^2 \tag{4.23}$$

Then, the constraint of the minimum turning radius is defined as follows.

$$\varrho \leq \frac{1}{R_{min}} \tag{4.24}$$

where R_{min} denotes the minimum turning radius.

The constraint of the safe distance between AV and obstacles can be expressed as

$$(X - X_{obs})^2 + (Y - Y_{obs})^2 < d_{obs}^2 \tag{4.25}$$

where (X_{obs}, Y_{obs}) denotes the position coordinate of the obstacle, d_{obs} denotes the minimum safe distance.

Based on the nonlinear discrete system (4.16), NMPC is applied to the state prediction of AVs considering multiple sampling periods. Defining the prediction horizon and control horizon as N_p and N_c for the prediction process, where $N_p \geq N_c$.

Furthermore, the predictive outputs are derived as

$$\begin{aligned}
y(k+1) &= g[x(k+1), u(k+1)] \\
y(k+2) &= g[x(k+2), u(k+2)] \\
&\vdots \\
y(k+N_c) &= g[x(k+N_c), u(k+N_c)] \\
y(k+N_c+1) &= g[x(k+N_c+1), u(k+N_c)] \\
&\vdots \\
y(k+N_p) &= g[x(k+N_p), u(k+N_c)]
\end{aligned} \tag{4.26}$$

Then, the output and control sequences are obtained.

$$\mathbf{y}(k+1) = [y(k+1), y(k+2), \cdots, y(k+N_p)]^T \tag{4.27}$$

$$\mathbf{u}(k+1) = [u(k+1), u(k+2), \cdots, u(k+N_c)]^T \tag{4.28}$$

In the optimization process, NMPC focuses on tracking the reference sequence that is expressed as

$$\mathbf{y}_{ref}(k+1) = [y_{ref}(k+1), y_{ref}(k+2), \cdots, y_{ref}(k+N_p)]^T \tag{4.29}$$

For path optimization, it aims to minimize the tracking error between the reference and the actual output.

$$\min \|\mathbf{y}(k+1) - \mathbf{y}_{ref}(k+1)\| \tag{4.30}$$

Comprehensively considering the tracking error and the control energy of AVs, the cost function for path optimization at the time step k is proposed as follows.

$$\begin{aligned}
\Upsilon_{static}(k) = &\sum_{i=1}^{N_p} \{[y(k+i|k) - y_{ref}(k+i|k)]^T Q [y(k+i|k) - y_{ref}(k+i|k)]\} \\
&+ \sum_{i=1}^{N_c} \{u(k+i|k)^T R u(k+i|k)\}
\end{aligned} \tag{4.31}$$

where Q and R are weighting matrices. Q is used to describe the weighting of the tracking error $y(k+i|k) - y_{ref}(k+i|k)$. R is the weighting regarding the control vector $u(k+i|k)$. Both Q and R are diagonal matrices.

Based on the constructed cost function, the path optimization issue with NMPC is described as

$$\min_{\mathbf{u}(k)} \Upsilon_{satic}(k) \tag{4.32}$$

Additionally, (4.32) should be subject to multiple constraints shown as follows.

$$x(k+i|k) = x(k+i-1|k) + F[x(k+i-1|k), u(k+i-1|k)] \tag{4.33}$$

$$y(k+i-1|k) = g[x(k+i-1|k), u(k+i-1|k)] \tag{4.34}$$

$$u_{min} \le u(k+i|k) \le u_{max} \tag{4.35}$$

$$\Delta y_{min} \le y(k+i|k) - y_{ref}(k+i|k) \le \Delta y_{max} \tag{4.36}$$

$$\varrho(k+i|k) \le 1/R_{min} \tag{4.37}$$

$$(X_{predict}(i) - X_{obs}(i))^2 + (Y_{predict}(i) - Y_{obs}(i))^2 < d_{obs}^2 \tag{4.38}$$

By solving the above optimization issue at the time step k, the optimal control sequence can be worked out.

$$\mathbf{u}(k+1|k) = [u(k+1|k), u(k+2|k), \cdots, u(k+N_c|k)]^T \tag{4.39}$$

It can be found that $\mathbf{u}(k+1|k)$ consists of N_c control vectors. $u(k+1|k)$ is usually used to calculate the planned path.

The above optimization issue is conducted at the time step k. Next step is the path optimization at the time step $k+1$, which has the similar solving process. Since the optimization issue is a nonlinear optimization with multi constraints, the sequential quadratic programming (SQP) method is used with the Matlab software [15].

Since too many constraints are considered in the optimization issue, it may cause the phenomenon of no solution. We defined the following rules. The first step is to enlarge the boundaries of tracking errors. If it does not work, then taking the second step, i.e., decreasing the velocity v. With the above two step, we can find the feasible solution to this optimization issue.

4.3.3 Path Planning for Moving Obstacles

4.3.3.1 Trajectory Prediction for Moving Obstacles

Since the motion of moving obstacles usually has uncertainties, it brings a challenge to the dynamic path planning. Accurate motion prediction of moving obstacles is in favor of the performance advancement of the dynamic path planning algorithm. In this subsection, a data-driven prediction algorithm is proposed via the polynomial fitting.

The moving trajectories of obstacles can be obtained by sensors, e.g., camera, GPS, radar, etc. Assuming the sampling time is T_{obs}, N sampling points at the time step k are used for trajectory prediction. Then, the position coordinates of N sampling points are expressed as

$$\mathbf{X}_{obs}(k) = [X_{obs}(k-N+1), X_{obs}(k-N+2), \cdots, X_{obs}(k)]^T \tag{4.40}$$

$$\mathbf{Y}_{obs}(k) = [Y_{obs}(k-N+1), Y_{obs}(k-N+2), \cdots, Y_{obs}(k)]^T \tag{4.41}$$

Considering the smoothness of the predicted path, the fifth degree polynomial is used to predict the moving trajectory of the obstacle vehicle. The fitting trajectory can be written as follows.

$$Y_{obs}^{fit} = \alpha_0 + \alpha_1 X_{obs} + \alpha_2 X_{obs}^2 + \alpha_3 X_{obs}^3 + \alpha_4 X_{obs}^4 + \alpha_5 X_{obs}^5 \tag{4.42}$$

where $\alpha_i (i = 0, 1, 2, 3, 4, 5)$ denotes the polynomial coefficient of the fifth degree polynomial.

Then, the least square method is used to solve the polynomial coefficients. The issue to solve the polynomial coefficients can be transformed into minimizing the error between the fitting value Y_{obs}^{fit} and the actual value Y_{obs}. It yields that

$$\min \|\Lambda(k)\alpha(k) - \mathbf{Y}_{obs}(k)\|_2^2 \tag{4.43}$$

where fitting coefficient matrix $\Lambda(k)$ and polynomial coefficient vector $\alpha(k)$ are derived by

$$\Lambda(k) = \begin{bmatrix} 1 & X_{obs}(k-N+1) & \cdots & X_{obs}(k-N+1)^5 \\ 1 & X_{obs}(k-N+2) & \cdots & X_{obs}(k-N+2)^5 \\ \vdots & \vdots & \vdots & \vdots \\ 1 & X_{obs}(k) & \cdots & X_{obs}(k)^5 \end{bmatrix}_{N\times 6} \tag{4.44}$$

$$\alpha(k) = [\alpha_0(k), \alpha_1(k), \alpha_2(k), \alpha_3(k), \alpha_4(k), \alpha_5(k)]^T \tag{4.45}$$

By solving the quadratic programming issue, the polynomial coefficient vector $\alpha(k)$ can be figured out. Then, the fitted fifth degree polynomial function can be applied to the trajectory prediction.

Based on (4.42), the slope of the fifth degree polynomial at the position $X_{obs}(k)$ is derived as

$$\begin{aligned} K(k) = \alpha_1(k) + 2\alpha_2(k)X_{obs}(k) + 3\alpha_3(k)X_{obs}^2(k) \\ +4\alpha_4(k)X_{obs}^3(k) + 5\alpha_5(k)X_{obs}^4(k) \end{aligned} \tag{4.46}$$

Moreover, the yaw angle of the obstacle vehicle at the position $X_{obs}(k)$ is derived as

$$\varphi_{obs}(k) = \arctan K(k) \tag{4.47}$$

Besides, the velocity of the obstacle vehicle is simplified as the average velocity within the sampling period.

$$v_{obs}(k) = \sqrt{(X_{obs}(k) - X_{obs}(k-1))^2 + (Y_{obs}(k) - Y_{obs}(k-1))^2} / T_{obs} \tag{4.48}$$

The final step is to work out the prediction position. With the assumption of constat velocity of the obstacle vehicle within the prediction horizon N_{obsp}, the predicted position of the obstacle vehicle is derived as follows.

$$X_{obs}(k+1|k) = X_{obs}(k) + T_{obs}v_{obs}(k)\cos\varphi_{obs}(k) \tag{4.49}$$

$$\begin{aligned} Y_{obs}(k+1|k) = \alpha_0(k) + \alpha_1(k)X_{obs}(k+1|k) + \alpha_2(k)X^2_{obs}(k+1|k) \\ +\alpha_3(k)X^3_{obs}(k+1|k) + \alpha_4(k)X^4_{obs}(k+1|k) + \alpha_5(k)X^5_{obs}(k+1|k) \end{aligned} \tag{4.50}$$

The trajectory prediction at the next time step $k+1$ has the similar solving process.

4.3.3.2 NMPC for Path Optimization

Based on the predicted trajectory of the obstacle vehicle, the following cost function is constructed for dynamic path planning to avoid the moving obstacle.

$$\Upsilon_{moving}(k) = \sum_{i=1}^{N_p}\{D(k+i|k)^T QD(k+i|k)\} + \sum_{i=1}^{N_c}\{u(k+i|k)^T Ru(k+i|k)\} \tag{4.51}$$

where $D(k+i|k)$ denotes the distance between the planned path of AV and the predicted trajectory of the obstacle.

Combining the static path planning and the dynamic path planning, the integrated path planning issue can be described as

$$\min_{\mathbf{u}(k)}[\Upsilon_{satic}(k) + \sum_{i=1}^{n}\Upsilon^i_{moving}(k)] \tag{4.52}$$

where $\Upsilon^i_{moving}(k)$ denotes the cost function of ith moving obstacle.

The integrated optimization issue should be subject to the following constraints:

$$x(k+i|k) = x(k+i-1|k) + F[x(k+i-1|k), u(k+i-1|k)] \tag{4.53}$$

$$y(k+i-1|k) = g[x(k+i-1|k), u(k+i-1|k)] \tag{4.54}$$

$$u_{min} \leq u(k+i|k) \leq u_{max} \tag{4.55}$$

$$\Delta y_{min} \leq y(k+i|k) - y_{ref}(k+i|k) \leq \Delta y_{max} \tag{4.56}$$

$$\varrho(k+i|k) \leq 1/R_{min} \tag{4.57}$$

$$(X_{predict}(i) - X_{obs}(i))^2 + (Y_{predict}(i) - Y_{obs}(i))^2 < d^2_{obs} \tag{4.58}$$

Finally, the control vector $u(k+1|k)$ is used to calculate the planned path, which comprehensively considers the static and dynamic planning.

4.3.4 Simulation and Validation

To evaluate the performance of the designed path planning algorithm, three test cases are designed and carried out in the Matlab software. Both static and dynamic obstacles are considered in the simulation case. Since the simulation scenarios are high-speed driving condition, the safe distance is set relatively large to guarantee the safety when conducting the collision avoidance. If the driving scenarios is low-speed condition, the safe distance can be decreased.

4.3.4.1 Case Study 1

The first test case aims to verify the performance of the path planning algorithm addressing the static obstacles. The position coordinates of the start point S and the final point F are set as (0,0) and (600,0), respectively. There exist 11 static obstacles in the closed area, and the AV must go across these obstacles from the start point S to the final point F.

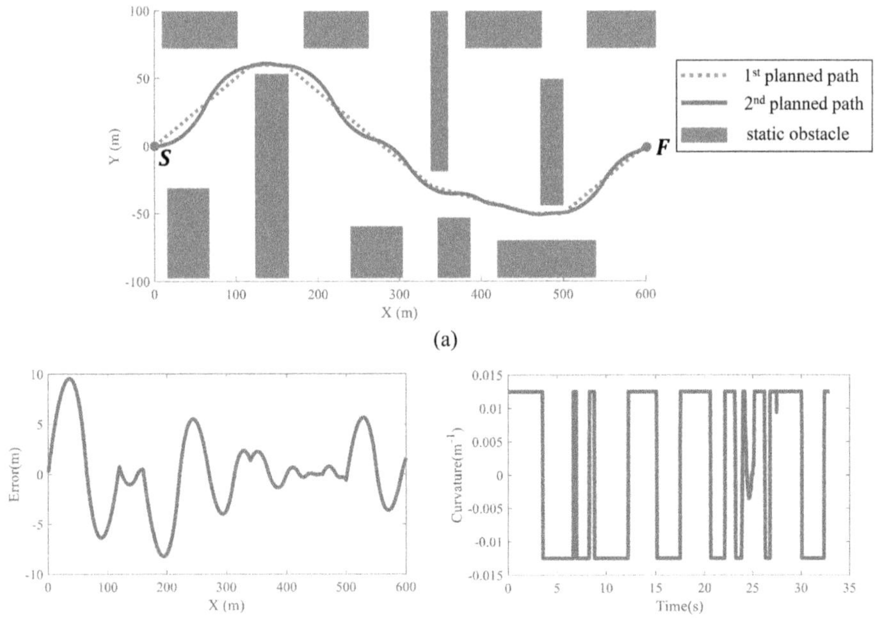

FIGURE 4.18: Simulation results in static obstacles condition: (a) planned path; (b) error; (c) curvature.

The test results are illustrated in Figure 4.18. Figure 4.18(a) shows the path planning results. We can find that the first planned path is made up of a set of line segments, which is planned by the visibility graph method.

Although it is the shortest path from the start point S to the final point F, it is unacceptable for path tracking control due to the large curvature at the turning point. Driving on a large curvature road with high speed tends to cause the loss of stability, which is harmful to driving safety. We can find that the second planned path by NMPC is smoother than the first planned path by the visibility graph method. Meanwhile, the safe distance between the AV and the static obstacle can be guaranteed. The error between the second planned path and the first planned path is shown in Figure 4.18(b).

The large errors usually exist near the turning points due to enlarging the turning radius. Moreover, Figure 4.18(c) shows the curvature change of the second planned path with NMPC. We can find that the path curvature is always within the control boundaries. Namely, the minimum turning radius of the vehicle is limited by 80 m. Based on the above analysis, it can be concluded that the proposed path planning algorithm is effective to realize collision avoidance of multiple static obstacles. Safety and travel efficiency are considered in the path planning process.

4.3.4.2 Case Study 2

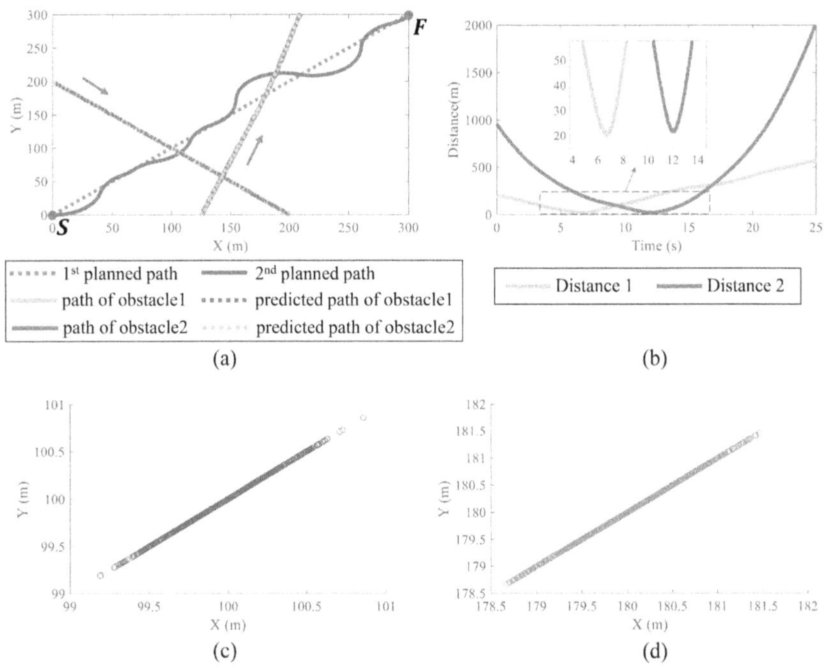

FIGURE 4.19: Simulation results in moving obstacles condition: (a) planned path; (b) distance between vehicles; (c) predictive collision point 1; (d) predictive collision point 2.

The second test case aims to evaluate the performance of the path planning algorithm addressing the moving obstacles. For the AV, the position coordinates of the start point S and the final point F are set as (0,0) and (300,300), respectively. Other two moving obstacle vehicles come from two different directions. The speed and path are unknown for the AV.

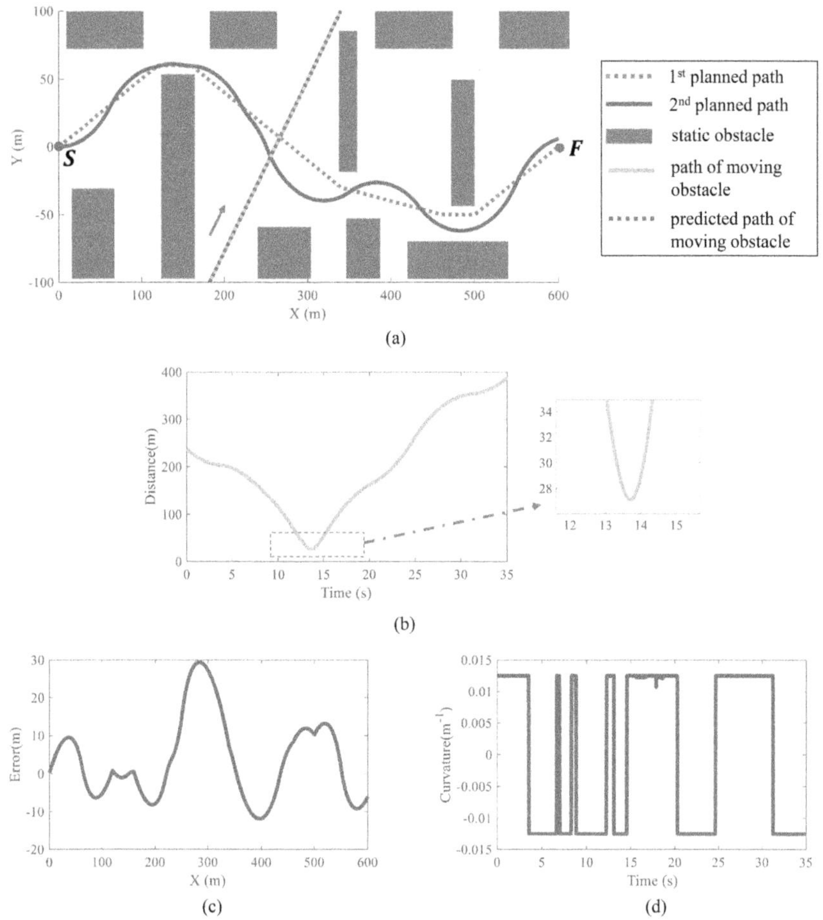

FIGURE 4.20: Simulation results in mixed condition: (a) planned path; (b) distance between vehicles; (c) error; (d) curvature.

The simulation results are illustrated in Figure 4.19. Figure 4.19(a) shows the path planning result. We can find that the planned path has smaller curvature near the conflict point, which aims to enlarge the safe distance between the AV and the obstacle vehicle. Moreover, the predicted path has a high degree of coincidence with the actual trajectory, which indicates that the trajectory prediction algorithm has high prediction accuracy. It can favor the collision avoidance of moving obstacles. To further analysis the quality of the planned path, the relative distances between the AV and the obstacles are shown in Figure 4.19(b). It can be found that the minimum safe distance is about 20m, which is within the control boundaries, which means the planned path is much safe. The predicted collision point between the AV and the obstacle vehicle 1 is shown in Figure 4.19(c), and the predicted collision point between the AV and the obstacle vehicle 2 is shown in Figure 4.19(d). With the position change of vehicles, the predicted collision points are not fixed. The predicted collision points are used to advance the accuracy of the path planning algorithm. The simulation results indicate that the path planning algorithm can realize collision avoidance of multiple moving obstacles.

4.3.4.3 Case Study 3

In the third test case, a mixed planning scenario is designed, which includes the static and moving obstacles. It aims to verify the path planning ability to avoid static and moving obstacles at the same time. For the AV, the position coordinates of the start point S and the final point F are set as (0,0) and (600,0), respectively. The speed and path of the obstacle vehicle are unknown for the AV.

The test results are displayed in Figure 4.20. The planned path is shown in Figure 4.20(a), from which we can find that the first planned path by the visibility graph method cannot realize dynamic collision avoidance. However, the second planned path by NMPC can realize static and dynamic collision avoidance at the same time. The distance between the AV and the obstacle vehicle is illustrated in Figure 4.20(b). The minimum distance is larger than 25m, which is within the safe distance boundary. The error between the first planned path and the second planned path is shown in Figure 4.20(c). The maximum error exists near the collision point. Besides, the curvature change of the second planned path with NMPC is displayed in Figure 4.20(d). We can find that the path curvature is always within the control boundaries. Based on the test results, it can be concluded that the path planning algorithm is effective realize static and dynamic collision avoidance at the same time.

4.3.5 Summary

In this section, an integrated path planning algorithm is designed, which can realize static and dynamic collision avoidance. Before designing the path planning algorithm, the pretreatment of driving scenarios is conducted, e.g., merging obstacles, and extending the boundaries of obstacles. Then, the visibility graph method is applied to the global path planning to avoid static obstacles. To improve the quality of the path planned by the visibility graph method, all kinds of constraints are taken into consideration, and NMPC is used to conduct the second path planning. To address the collision avoidance of moving obstacles, the data-driven approach with the polynomial fitting is used to predict the moving trajectories of moving obstacles. Furthermore, the path planning issue of static and dynamic collision avoidance is transformed into a nonlinear optimization with multiple constraints. NMPC is used to solve the optimization issue. Finally, three test cases are carried out to verify the feasibility and effectiveness of the proposed path planning algorithm. The results shows that the comprehensive path planning algorithm can provide the safe path for AVs. Since the lateral stability issue is considered in the path planning issue, the planned path is acceptable for path tracking control.

4.4 Path Tracking Control of AVs

This section presents the path tracking controller design for AVs. Linear-time-varying (LTV) MPC is applied for the integrated controller design, which takes the improvement of handling stability and path tracking performance into considerations based on the LTV vehicle model. Taking all kinds of constraints into account, including control vector constraints, lateral stability constraints, rollover prevention constraints and path tracking error constraints, the integrated controller is designed and worked out by solving the optimization problem with multi-constraints. To evaluate the performance of the integrated controller, the double-lane change maneuver and the sinusoidal path maneuver are carried out.

4.4.1 Linearized and Discretized Model for Path Tracking Control

To realize path tracking control with the LTV MPC approach, the 3 DoF nonlinear vehicle dynamic model is rewritten together as

$$\dot{\beta} = -r - \frac{m_s h_s}{m v_x}\ddot{\phi} + \frac{1}{m v_x}\Sigma F_y \tag{4.59}$$

$$\dot{r} = \frac{I_{xz}}{I_z}\ddot{\phi} + \frac{1}{I_z}\Sigma M_z \tag{4.60}$$

$$\dot{\varphi} = r \tag{4.61}$$

$$\ddot{\phi} = \frac{I_{xz}}{I_x}\dot{r} + \frac{1}{I_x}\Sigma L_x \tag{4.62}$$

$$\dot{Y} = v_x \sin\varphi + v_y \cos\varphi \tag{4.63}$$

To simplify the controller design procedure, two front wheels and two rear wheels can be assumed to be lumped together at the centers of front and rear axles, respectively. As a result, the four-wheel vehicle model can be simplified as a single track model.

Furthermore, (4.59) to (4.63) are written in the state-space form, the state vector and the control vector are defined as $x = [\beta, r, \varphi, \phi, \dot{\phi}, Y]^T$ and $u = \delta_f$, respectively.

$$\dot{x}(t) = f(x(t), u(t)) \tag{4.64}$$

Conducting Taylor expansion at the point (x_t, u_t), (4.64) is expressed with an approximated LTV system.

$$\dot{x}(t) = A_t x(t) + B_t u(t) + d_t \tag{4.65}$$

where d_t denotes the linearization residual item, the time-varying coefficient matrices A_t and B_t are given by

$$A_t = \frac{\partial f}{\partial x}|_{x_t, u_t}, \quad B_t = \frac{\partial f}{\partial u}|_{x_t, u_t} \tag{4.66}$$

For controller design, (4.65) is discretized as

$$\begin{cases} x(k+1) = A(k)x(k) + B(k)u(k) + d(k) \\ y(k) = C(k)x(k) \\ u(k) = u(k-1) + \Delta u(k) \end{cases} \tag{4.67}$$

where $A(k) = I + A_t T$, $B(k) = B_t T$, $C(k) = I_6 \in I^{6\times 6}$, $d(k) = d_t T$ and T is the sampling time.

4.4.2 Integrated Controller Design

4.4.2.1 Control System Framework

The integrated control framework for AV is shown in Figure 4.21. It can be found from the figure that the control system framework mainly consists of the integrated controller and the longitudinal controller. The longitudinal motion control algorithm is not the key point of this work, so it will not be introduced in this section. This work mainly focuses on the integrated controller design, which includes the path tracking control, handling stability control and rollover prevention control. The steering angle allocation algorithm is designed based on the Ackerman steering geometry. The error generator aims to calculate the tracking errors between the actual values and the references. The desired sideslip angle β_d and the desired yaw rate r_d come from the reference model. Besides, the desired Y_d and φ_d are provided by the target path.

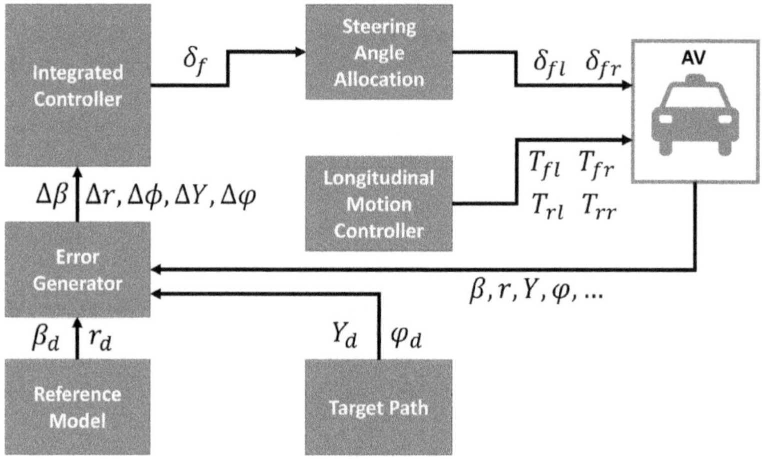

FIGURE 4.21: Integrated control framework for AV.

4.4.2.2 Handling Stability Improvement

The handling performance of vehicles is reflected by the yaw rate. The desired yaw rate is related to the front steering angle, which can be expressed as [72].

$$\hat{r}_d = \frac{K_r}{1 + \tau_r s}\delta_f \tag{4.68}$$

$$K_r = \frac{v_x}{l(1 + \zeta v_x^2)} \tag{4.69}$$

$$\zeta = \frac{m}{l^2}(\frac{l_f}{k_r} - \frac{l_r}{k_f}) \tag{4.70}$$

where τ_r is the time constant, K_r is the gain of desired yaw rate to the steering angle of front wheel.

Providing that the limit of the yaw rate is the steady-state yaw rate associated with the maximum lateral tire forces, as a result, the yaw rate limit can be defined as [11]

$$r_{\max} = \begin{cases} \frac{1+l_r/l_f}{mv_x}F_{yr\max}, & F_{yf\max} \geq \frac{lr}{lf}F_{yr\max} \\ \frac{1+l_f/l_r}{mv_x}F_{yf\max}, & F_{yf\max} < \frac{lr}{lf}F_{yr\max} \end{cases} \tag{4.71}$$

where $F_{yf\max}$ and $F_{yr\max}$ are the maximum lateral forces on front and rear axles, which can be expressed as

$$F_{yf\max} = \frac{l_r}{l}mg\mu_y \tag{4.72}$$

$$F_{yr\max} = \frac{l_f}{l} mg\mu_y \tag{4.73}$$

where μ_y is the lateral friction coefficient.

Substitution of (4.73) into (4.72) yields the following yaw rate limit.

$$|r| \leq r_{\max} = \frac{g\mu_y}{v_x} \tag{4.74}$$

Considering the limit of yaw rate, the desired yaw rate is reexpressed as

$$r_d = sign(\delta_f) \cdot \min(r_{\max}, |\hat{r}_d|) \tag{4.75}$$

The lateral stability of vehicles is usually reflected by the sideslip angle. Smaller sideslip angle means better lateral stability. Therefore, the desired sideslip angle is usually set as zero, i.e.,

$$\beta_d = 0 \tag{4.76}$$

The sideslip angle is constrained by the tire-ground adhesion condition with the following empirical formula [273].

$$|\beta| \leq \arctan(0.02\mu g) \tag{4.77}$$

Besides, $\beta - \dot{\beta}$ phase plane is usually applied to describing the vehicle lateral stability, i.e.,

$$|\beta + B_1\dot{\beta}| \leq B_2 \tag{4.78}$$

which describes the boundaries of vehicle stability margin, B_1 and B_2 are the boundary coefficients with respect to v_x and μ [233].

4.4.2.3 Rollover Prevention

The roll angle and its derivative with respect to time are usually used to evaluate the rollover prevention. The desired roll angle and its derivative are given by

$$\phi_d = 0, \quad \dot{\phi}_d = 0 \tag{4.79}$$

The boundary of roll angle is expressed with the empirical formula [8]

$$|\phi| \leq \phi_{\max} = \frac{Bm_s}{2(k_\phi - m_s g h_s)} \tag{4.80}$$

For rollover prevention, the phase plane method is used again, which is defined as follows [8].

$$|C_1\phi + C_2\dot{\phi}| \leq C_3 \tag{4.81}$$

where C_1 and C_2 are related to vehicle parameters, C_3 is the rollover index within the acceptable region.

4.4.2.4 Path Tracking Performance

The path tracking performance of AVs can be evaluated based on the lateral offset ΔY and yaw angle error $\Delta\varphi$. The desired lateral offset and yaw angle error are given by

$$\Delta Y_d = 0, \quad \Delta\varphi_d = 0 \tag{4.82}$$

The control boundary of the lateral offset ΔY is written as

$$|\Delta Y| < 0.5W_{road} - 0.5W_{AV} \tag{4.83}$$

where W_{road} is the width of the road and W_{AV} is the width of AV.

Besides, the control boundary of the yaw angle error $\Delta\varphi$ is defined by.

$$|\Delta\varphi| < \Delta\varphi_{\max} \tag{4.84}$$

where $\Delta\varphi_{\max}$ denotes the maximum yaw angle error.

4.4.2.5 LTV-MPC for Integrated Control

For LTV-MPC design, a new state vector is defined that includes the original state vector and control vector.

$$\xi(k) = [x(k), u(k-1)]^T \tag{4.85}$$

Furthermore, a new discrete state-space form of (4.67) can be derived as

$$\begin{cases} \xi(k+1) = \tilde{A}_k\xi(k) + \tilde{B}_k\Delta u(k) + \tilde{D}_k d(k) \\ y(k) = \tilde{C}_k\xi(k) \end{cases} \tag{4.86}$$

where $\tilde{A}_k = \begin{bmatrix} A(k) & B(k) \\ 0_{1\times6} & I \end{bmatrix}$, $\tilde{B}_k = \begin{bmatrix} B(k) \\ I \end{bmatrix}$, $\tilde{C}_k = \begin{bmatrix} I_6 \\ 0_{1\times6} \end{bmatrix}^T$ and $\tilde{D} = [I_6 \quad 0_{6\times1}]^T$,

Defining the predictive horizon N_p and the control horizon N_c, $N_p \geq N_c$, at the time step k, the state vectors during the prediction horizon can be expressed as

$$\xi(k+1|k), \xi(k+2|k), \cdots, \xi(k+N_p|k) \tag{4.87}$$

Besides, the control vectors during the control horizon can be written as

$$\Delta u(k|k), \Delta u(k+1|k), \cdots, \Delta u(k+N_c-1|k) \tag{4.88}$$

At the time step k, if the state vector $\xi(k)$, the control vector $\Delta u(k)$ and coefficient matrices i.e., $\tilde{A}_{p,k}$, $\tilde{B}_{p,k}$, $\tilde{C}_{p,k}$, and $\tilde{D}_{p,k}$ are known, the predicted state vectors can be derived as follows.

$$\begin{cases} \xi(p+1|k) = \tilde{A}_{p,k}\xi(p|k) + \tilde{B}_{p,k}\Delta u(p|k) + \tilde{D}_{p,k}d(p|k) \\ y(p|k) = \tilde{C}_{p,k}\xi(p|k) \end{cases} \tag{4.89}$$

where $p = k, k+1, \cdots, k+N_p-1$.

To simplify the design of LTV-MPC, assuming that $\tilde{A}_{p,k} = \tilde{A}_k$, $\tilde{B}_{p,k} = \tilde{B}_k$, $\tilde{C}_{p,k} = \tilde{C}_k$ and $\tilde{D}_{p,k} = \tilde{D}_k$, it yields that

$$\begin{aligned}
\xi(k+1|k) &= \tilde{A}_k\xi(k|k) + \tilde{B}_k\Delta u(k|k) + \tilde{D}_k d(k|k) \\
\xi(k+2|k) &= \tilde{A}_k^2\xi(k|k) + \tilde{A}_k\tilde{B}_k\Delta u(k|k) + \tilde{B}_k\Delta u(k+1|k) \\
&\quad + \tilde{A}_k\tilde{D}_k d(k|k) + \tilde{D}_k d(k+1|k) \\
&\vdots \\
\xi(k+N_c|k) &= \tilde{A}_k^{N_c}\xi(k|k) + \tilde{A}_k^{N_c-1}\tilde{B}_k\Delta u(k|k) + \cdots \\
&\quad + \tilde{B}_k\Delta u(k+N_c-1|k) + \tilde{A}_k^{N_c-1}\tilde{D}_k d(k|k) + \cdots \\
&\quad + \tilde{D}_k d(k+N_c-1|k) \\
&\vdots \\
\xi(k+N_p|k) &= \tilde{A}_k^{N_p}\xi(k|k) + \tilde{A}_k^{N_p-1}\tilde{B}_k\Delta u(k|k) + \cdots \\
&\quad + \tilde{A}_k^{N_p-N_c}\tilde{B}_k\Delta u(k+N_c-1|k) + \tilde{A}_k^{N_p-1}\tilde{D}_k d(k|k) + \cdots \\
&\quad + \tilde{D}_k d(k+N_p-1|k)
\end{aligned} \tag{4.90}$$

Combining the output vectors from $y(k+1|k)$ to $y(k+N_p|k)$ yields the output vector sequence.

$$\Upsilon(k) = [y^T(k+1|k), y^T(k+2|k), \cdots, y^T(k+N_p|k)]^T \tag{4.91}$$

Moreover, it can be expressed as

$$\Upsilon(k) = \bar{C}\xi(k|k) + \bar{E}\Delta U(k) + \bar{D}W(k) \tag{4.92}$$

where

$$\Delta U(k) = [\Delta u^T(k|k), \Delta u^T(k+1|k), \cdots, \Delta u^T(k+N_c-1|k)]^T \tag{4.93}$$

$$W(k) = [d^T(k|k), d^T(k+1|k), \cdots, d^T(k+N_p-1|k)]^T \tag{4.94}$$

$$\bar{C} = [(\tilde{C}_k\tilde{A}_k)^T, (\tilde{C}_k\tilde{A}_k^2)^T, \cdots, (\tilde{C}_k\tilde{A}_k^{N_p})^T]^T \tag{4.95}$$

$$\bar{E} = \begin{bmatrix}
\tilde{C}_k\tilde{B}_k & 0 & 0 & 0 \\
\vdots & \vdots & \vdots & \vdots \\
\tilde{C}_k\tilde{A}_k^{N_c-1}\tilde{B}_k & \cdots & \tilde{C}_k\tilde{A}_k\tilde{B}_k & \tilde{C}_k\tilde{B}_k \\
\vdots & \vdots & \vdots & \vdots \\
\tilde{C}_k\tilde{A}_k^{N_p-1}\tilde{B}_k & \cdots & \tilde{C}_k\tilde{A}_k^{N_p-N_c+1}\tilde{B}_k & \tilde{C}_k\tilde{A}_k^{N_p-N_c}\tilde{B}_k
\end{bmatrix} \tag{4.96}$$

$$\bar{D} = \begin{bmatrix} \tilde{C}_k\tilde{D}_k & 0 & 0 & 0 & 0 \\ \vdots & \vdots & \vdots & \vdots & \vdots \\ \tilde{C}_k\tilde{A}_k^{N_c-1}\tilde{D}_k & \tilde{C}_k\tilde{A}_k^{N_c-2}\tilde{D}_k & \cdots & 0 & 0 \\ \vdots & \vdots & \vdots & \vdots & \vdots \\ \tilde{C}_k\tilde{A}_k^{N_p-1}\tilde{D}_k & \tilde{C}_k\tilde{A}_k^{N_p-2}\tilde{D}_k & \cdots & \tilde{C}_k\tilde{A}_k\tilde{D}_k & \tilde{C}_k\tilde{D}_k \end{bmatrix} \tag{4.97}$$

Defining the reference vector sequence as

$$\Upsilon_r(k) = [y_r^T(k+1|k), y_r^T(k+2|k), \cdots, y_r^T(k+N_p|k)]^T \tag{4.98}$$

LTV-MPC design can be described as a general optimization problem, which aims to find the optimal control vector sequence ΔU so as to minimize the error between the predicted output vector sequence and reference vector sequence. To this end, the system cost function is defined as

$$J = (\Upsilon(k) - \Upsilon_r(k))^T Q(\Upsilon(k) - \Upsilon_r(k)) + \Delta U(k)^T R \Delta U(k) \tag{4.99}$$

where Q and R are diagonal weighting matrices to balance the tracking accuracy and control increment, respectively.

Therefore, the optimization of LTV-MPC can be expressed as

$$\min_{\Delta U(k)} J(k) \tag{4.100}$$

Subject to

$$\xi(p+1|k) = \tilde{A}_k\xi(p|k) + \tilde{B}_k\Delta u(p|k) + \tilde{D}_k d(p|k) \tag{4.101}$$

$$y(p+1|k) = \tilde{C}_k\xi(p+1|k) \tag{4.102}$$

where $p = k, k+1, \cdots, k+N_p-1$.

Meanwhile, the following constraints should be taken into consideration.

(1) Control vector constraints. They are constrained by the limit of the actuators, which is described as

$$\Delta u_{\min} \le \Delta u(p|k) \le \Delta u_{\max} \tag{4.103}$$

$$u_{\min} \le u(p|k) \le u_{\max} \tag{4.104}$$

where $p = k, k+1, \cdots, k+N_c-1$.

(2) Rollover prevention constraint, i.e., (4.80) and (4.81).

(3) Handling stability constraints, i.e., (4.74), (4.77) and (4.78).

(4) Path tracking constraints, i.e., (4.83) and (4.84).

After solving the aforementioned optimization problem with multi-constraint, the first element of $\Delta U(k)$ will be used to calculate the controller output at the time step k.

$$u(k|k) = u(k-1|k-1) + \Delta u(k|k) \tag{4.105}$$

At the next time step $k+1$, a new optimization is solved over a shifted prediction horizon again with the updated state $\xi(k+1)$.

In the solving process of LTV-MPC, it brings a challenge to find a feasible solution at any moment due to the multiple constraints. To this end, the priority of constraints is defined, i.e., Constraint (1)>Constraint (2)> Constraint (3)> Constraint (4). Compared with the path tracking performance, rollover prevention and handling stability are more important. Therefore, they have higher priority. If there is no solution to this issue, two steps will be conducted. The first step is to extend the boundary of the constraints from the low priority to the high priority in the safe range. If there is still no solution after conducting the first step, the second step is to stop the vehicle.

4.4.2.6 Weighting Matrices for Control Objectives

Considering that AVs have different control focuses under different conditions, adaptive weighting matrices are utilized for active safety control. For instances, under the low-speed condition, rollover prevention and handling stability are not very important. AVs care more about the path tracking performance. Therefore, larger control weighting for path tracking control should be considered . However, under the high-speed condition, especially for the extreme conditions, rollover prevention and handling stability are very critical for driving safety. Larger control weighting for rollover prevention and handling stability control should be considered.

The weight of sideslip angle is zero when β is within the acceptable region, and it is a very large value when β approaches the limit. Hence, the weight of sideslip angle is defined by

$$q_\beta = \vartheta_\beta(|\beta| + (1 - \beta_{\max}))^{n_\beta} \tag{4.106}$$

where ϑ_β is a constant positive value and n_β is a sufficiently large positive number.

Besides, the weight of roll angle is defined by

$$q_\phi = \vartheta_\phi(|\phi| + (1 - \phi_{\max}))^{n_\phi} \tag{4.107}$$

where ϑ_ϕ is a constant positive value and n_ϕ is a sufficiently large positive number.

The weight of yaw rate tracking is set as a constant positive value $q_r = \vartheta_r$. Besides, the weights of yaw angle error and lateral offset are expressed as $q_\phi = \vartheta_\phi / \Delta\varphi_{\max}$ and $q_Y = \vartheta_Y / \Delta Y_{\max}$, where ϑ_ϕ and ϑ_Y are constant positive values.

4.4.3 Simulation and Analysis

To verify the performance of the proposed path-tracking controller, two test maneuvers are carried out, i.e., double-lane change (DLC) maneuver and sinusoidal path maneuver. All tests are conducted with the software Matlab Simulink.

4.4.3.1 Double-Lane Change Maneuver

In this test case, the AV is subjected to a DLC maneuver. DLC test is usually to evaluate the handling stability, especially for high-speed driving. For AVs, the test of DLC path tracking can not only evaluate the path tracking performance, but also test the handling stability. In the this case, the longitudinal velocity is set as 20m/s, and the road friction coefficient is set as 0.35, 0.55 and 0.85 to simulate the snowy road, wet road and dry road.

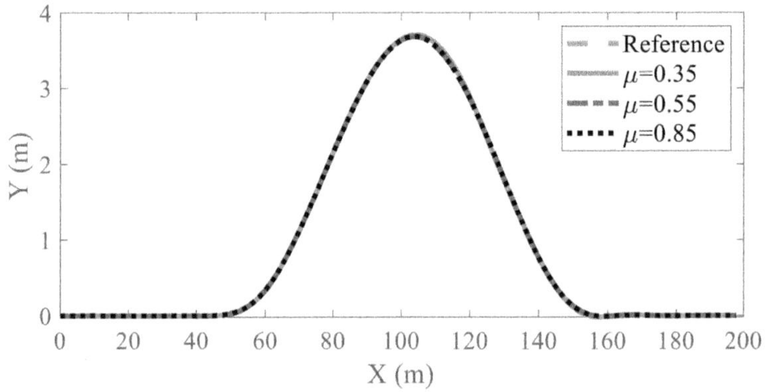

FIGURE 4.22: Path tracking results with different road friction coefficients.

The path tracking test result is illustrated in Figure 4.22. It can be found that the designed controller shows good path tracking performance on different roads. For detailed analysis, the path-tracking error is shown in Figure 4.23. On the snowy road, the maximum lateral offset is less than 0.06 m. With the increase of road friction coefficient, the path tracking error becomes smaller. It can also be concluded that the designed controller has good robustness against different road friction coefficients.

Moreover, sideslip angles and yaw rates in this test case are shown in Figures 4.24 and 4.25, respectively. We can see from Figures 4.24 that the designed controller can realize small sideslip angle even on the snowy road, which indicates the AV has good handling stability. Besides, front wheel steering angles are illustrated in Figure 4.26. It can be seen that the control output will rise with the decrease of road friction coefficient. On the low-adhesion coefficient road, to guarantee the lateral stability, AVs need conduct larger steering angle to provide the efficient lateral tire force.

4.4.3.2 Sinusoidal Path Maneuver

The second test case is a sinusoidal path maneuver, which aims to verify the superiority of LTV-MPC compared with linear static MPC (LS-MPC) [77]. The longitudinal velocity is set as 12 m/s and the road friction coefficient is set as 0.55 in this test. Compared with the DLC maneuver, the sinusoidal

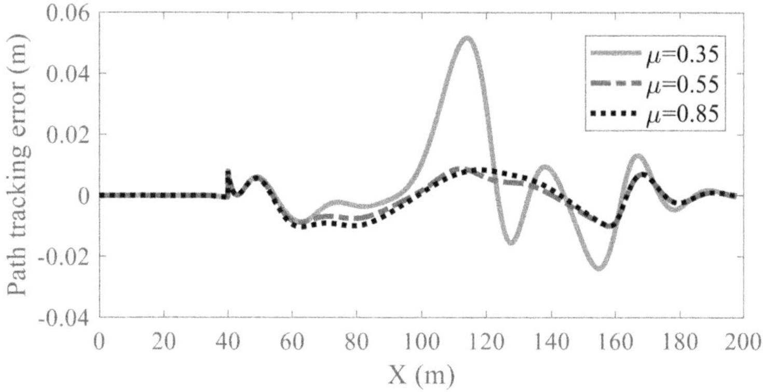

FIGURE 4.23: Path tracking errors with different road friction coefficients.

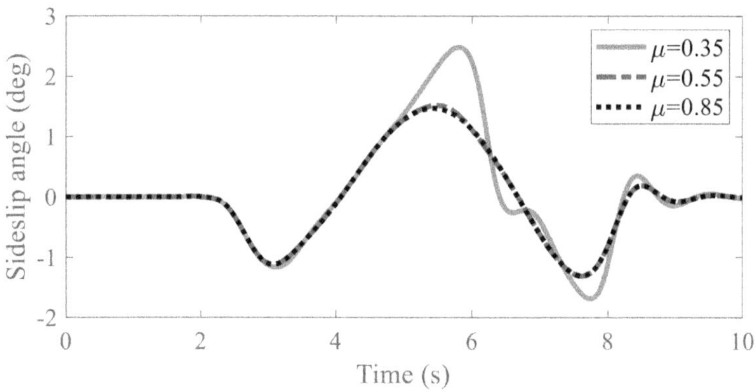

FIGURE 4.24: Sideslip angle with different road friction coefficients.

path maneuver has larger curvature. As a result, it will bring larger lateral acceleration, which is a challenge to the lateral stability of AVs.

The path tracking test result of sinusoidal path maneuver is depicted in Figure 4.27. It can be seen that LTV-MPC control algorithm shows better path tracking performance with smaller tracking error than LS-MPC algorithm. Moreover, LTV-MPC control algorithm has faster convergence rate. Figure 4.28 shows the analysis results of the path tracking error. We can find that the LTV-MPC algorithm always has smaller lateral offset than the LS-MPC algorithm. It indicates that LTV-MPC can deal with the large curvature path tracking problem better.

For handling stability analysis, sideslip angles and yaw rates of two control algorithms are illustrated in Figures 4.29 and 4.30, respectively. Since the curvature of sinusoidal path maneuver is much larger DLC maneuver, both

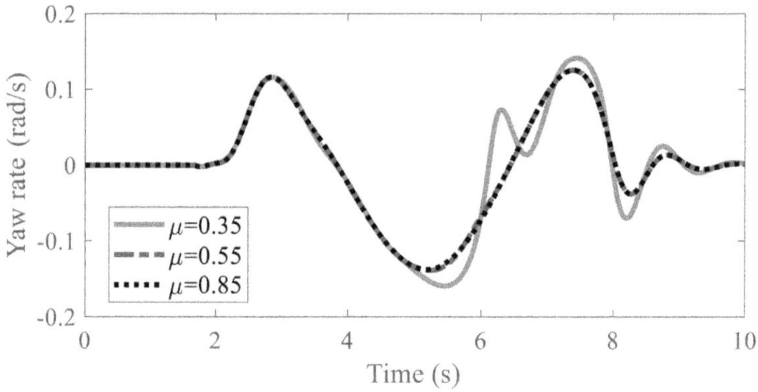

FIGURE 4.25: Yaw rate with different road friction coefficients.

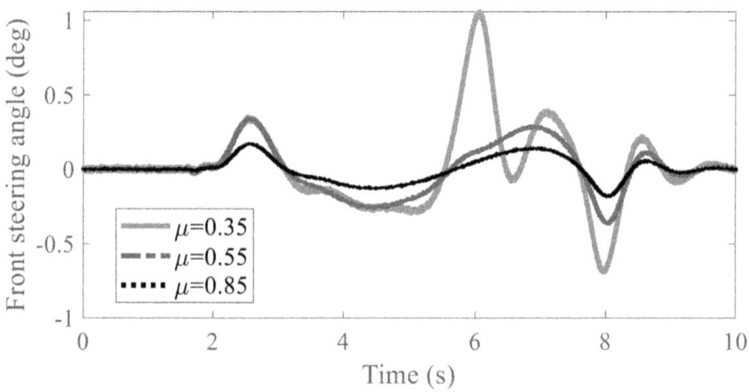

FIGURE 4.26: Front wheel steering angle with different road friction coefficients.

the LTV-MPC control algorithm and the LS-MPC control algorithm show large sideslip angles. By comparison, LTV-MPC has smaller sideslip angle. As to yaw rates, LTV-MPC has smaller tracking error as well. From the above test and analysis results, it can be concluded that LTV-MPC can bring better handling stability than LS-MPC.

The front wheel steering angles of two control algorithm are depicted in Figure 4.31. It can be found that LTV-MPC can obtain smaller tracking errors with smaller control outputs because the model of LTV-MPC is time-varying and more complex than that of LS-MPC. Since the model of LS-MPC is simpler, it cannot calculate accurate control outputs.

It can be concluded from the test results of the sinusoidal path maneuver that LTV-MPC control algorithm has better path tracking performance and

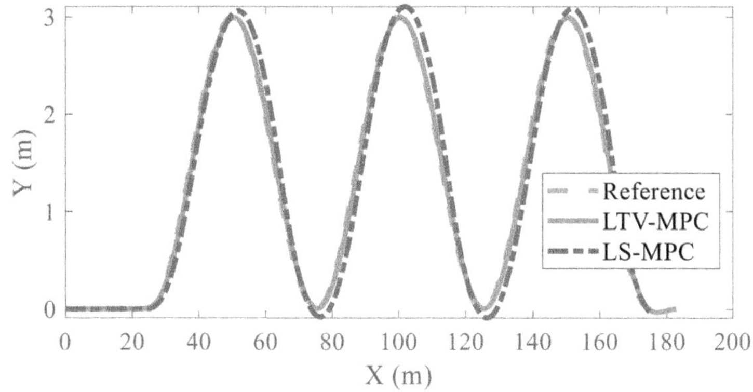

FIGURE 4.27: Path tracking results under sinusoidal path maneuver.

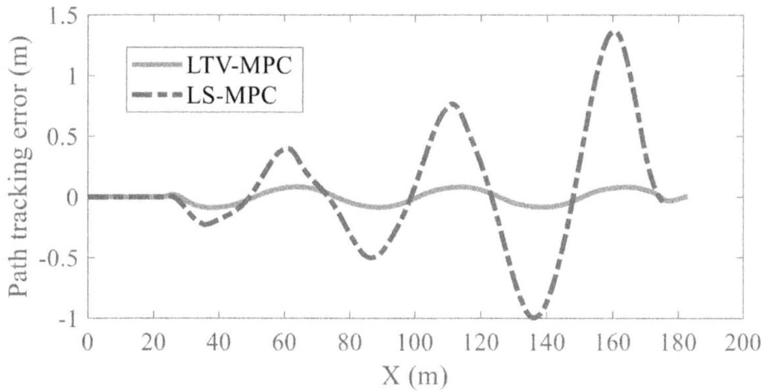

FIGURE 4.28: Path tracking errors under sinusoidal path maneuver.

handling stability than the LS-MPC control algorithm. In addition, LTV-MPC control algorithm shows good performance to deal with large curvature path tracking condition.

From the aforementioned tests and analysis results, we can conclude that the designed integrated controller can improve the AV's handling stability and path tracking performance very well. Meanwhile, it shows good robust performance to deal with large curvature path tracking condition and low road friction coefficient condition.

4.4.4 Summary

In this section, the path tracking control problem and the handling stability control problem of AVs are studied. To solve the aforementioned two

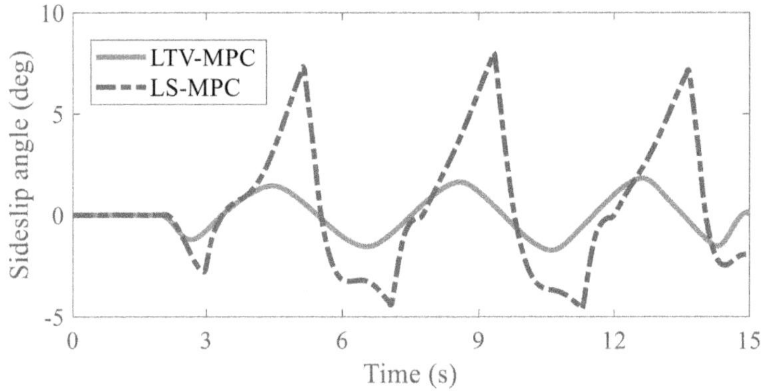

FIGURE 4.29: Sideslip angle under sinusoidal path maneuver.

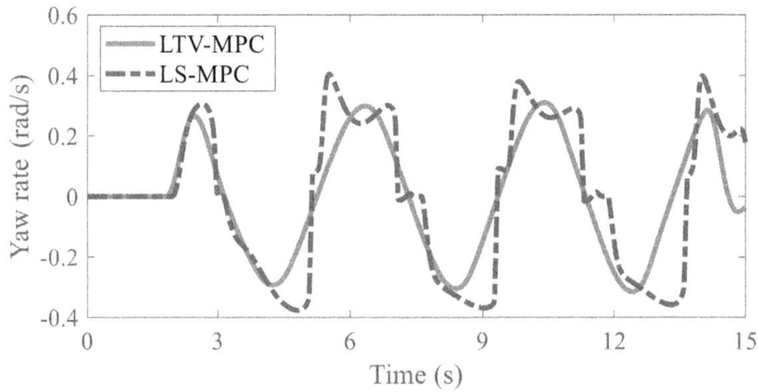

FIGURE 4.30: Yaw rate under sinusoidal path maneuver.

problems, an integrated controller is designed based on LTV-MPC control algorithm. To improve AVs' path tracking performance and handling stability simultaneously, the LTV-MPC control algorithm is applied. For the integrated controller design, the 3 DoF nonlinear vehicle dynamic model is built, linearized and discretized. In the LTV-MPC control algorithm, all kinds of constrains are taken into consideration including control vector constraints, lateral stability constraints, rollover prevention constraints and path tracking error constraints. By solving the optimization problem with multi-constraint, the integrated controller is worked out. To verify the performance of the designed integrated controller, two test conditions are carried out. The test results indicate that the LTV-MPC controller can improve AV's path tracking performance and handling stability remarkably. Moreover, compared with traditional LS-MPC, LTV-MPC shows better control performance. Finally, it can

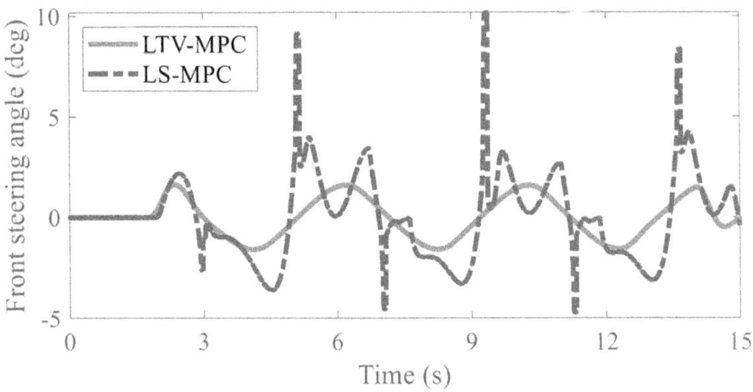

FIGURE 4.31: Front wheel steering angle under sinusoidal path maneuver.

be concluded that the integrated controller can improve AV's path tracking performance and handling stability simultaneously. Meanwhile, it has good robust performance to deal with low road friction coefficient condition and large curvature path tracking condition.

4.5 Conclusion

This chapter aims to deal with the motion planning and control of AVs. In the motion planning part, a human-like planner is designed for AVs to realize trajectory planning on highways. APF model is adopted to describe the social characterizations of vehicles and embedded in the motion-planning module. And MPC is used the speed and path prediction of the AV. Finally, the trajectory planning issue transformed into a closed-loop interactive optimization problem with multi-constraint. Three test cases are carried out to verify the proposed trajectory planning algorithm. Test results indicate that the trajectory planning algorithm is capable of planning human-like trajectory for AVs on highways, and adaptive to different driving styles of surrounding vehicles.

Besides, a comprehensive path planner is designed for AVs on unstructured roads, which can realize collision avoidance of both static and moving obstacles. The visibility graph method is used firstly to plan the global collision avoidance path across the static obstacles. Then, comprehensively considering the uncertain moving obstacles and multiple constraints, e.g., lateral stability, minimum turning radius, safety, etc., NMPC is used to optimize the path planned by the visibility graph method. In the optimizing process, model based polynomial fitting is utilized to predict the moving trajectories of uncertain moving obstacles. Simulation results show that the path planning algorithm can avoid both static and moving obstacles very well.

Finally, the LTV-MPC control algorithm is applied to the path tracking control of AVs. To guarantee the driving safety of AVs in extreme conditions, all kinds of active safety performance indexes and constrains are considered in the integrated controller design, including handling stability, rollover prevention, control capacity, and path tracking error. Then, the controller solving is transformed into a optimization problem with multi-constraint. Adaptive weighting is in favor of the balance between multiple performance indexes. Test results indicate that the LTV-MPC controller has better path tracking performance and handling stability than LS-MPC controller. Moreover, the proposed controller has good robustness to address extreme conditions, e.g., low road friction coefficient condition and large curvature path tracking condition.

Chapter 5

Human-Like Decision Making for Autonomous Vehicles with Noncooperative Game Theoretic Method

DOI: 10.1201/9781003287087-5

5.1 Background

As discussed in Chapter 1, AVs will be used for different purposes in the future, e.g., school buses, fire trucks, taxis, etc., which require personalized driving on different performances, e.g., safety, comfort, and efficiency. Besides, different passengers have individual travel demands. To this end, the concept of human-like decision making is proposed. Compared with the traditional model-based approach and the data-driven approach, the game theoretic approach is effective to formulate the human-like interaction and decision making [286]. The lane-change decision making is a common issue for AVs, which has been widely studied. Actually, the lane-change process reflects the game between multiple players. The noncooperative game theoretic approach is easy to describe the lane-change interaction and decision making. Due to the few players, the interaction process of the lane-change scenario is relatively simple.

To further study the human-like decision making of AVs in complex driving environment, the urban scenarios are usually investigated, e.g., unsignalized roundabouts. Unsignalized roundabout intersections are usually considered to be more complex and challenging than crossroad intersections, with respect to multi-vehicle interactions [174, 263, 43]. A roundabout is defined as a circular intersection in which all vehicles travel around a circular island at the center with the counter-clockwise direction (driving on the right) [3]. Compared to crossroad intersections, traffic lights are not necessary for roundabouts to control the traffic flow. Therefore, for the entering vehicles, a complete stop may not be required. As a result, the traffic delay can be decreased, and the traffic capacity can be improved [187]. In general, the traffic rules at roundabouts are defined as follows: (1) Circulating vehicles have the priority over entering and merging vehicles; (2) exiting vehicles have the priority over entering vehicles; and (3) large vehicles have the priority over small vehicles. The aforementioned rules are able to bring remarkable advances to the traffic efficiency and safety at roundabouts [176]. Although the above traffic rules can help reduce the traffic conflicts, with the increase of traffic flow, especially during peak hours, congestion and conflicts are inevitable due to different travel objectives, and personalized driving characteristics and driving behaviors of human drivers. Therefore, the complex unsignalized roundabout urban scenario is a great challenge for AVs. Human-like decision making is effective in dealing with the interaction and decision making of AVs in complex conditions.

In this chapter, the human-like decision-making algorithm is designed for AVs with the noncooperative game theoretic approach. Two typical driving scenarios, i.e., lane-change and unsignalized roundabout are studied for algorithm verification.

5.2 Human-Like Lane Change for AVs

In this study, to further advance the human-like decision-making algorithms for AVs, typical driving styles are defined at first to reflect different driving characteristics during the modeling phase. Beyond this, the decision-making problem of AVs is formulated using game theory. Two noncooperative game approaches, i.e., the Nash equilibrium and Stackelberg games, are adopted to address decision making and interactions for AVs. Finally, the developed human-like decision-making algorithms for AVs are tested and validated via simulation in various scenarios.

5.2.1 Problem Description and Human-Like Decision-Making Framework

5.2.1.1 Human-Like Lane Change Issue

A common lane-change scenario on a three-lane highway is shown in Figure 5.1. Ego car (EC) is an AV moving on the middle lane, and lead car 2 (LC2) moves in front of EC with a lower speed. Due to the lower speed of LC2, EC has to make a decision, i.e., decelerating and following LC2, or changing lanes to the left lane or the right lane. The decision-making result of EC is directly influenced by the motion states and driving behaviors of surrounding vehicles, especially by the adjacent cars (ACs), i.e., AC1 and AC2. For instance, if ACs are more aggressive than EC, ACs will not give ways to EC. As a result, EC must yield and decelerate to guarantee driving safety. Additionally, the different driving styles or characteristics of EC also have significant effects on the decision-making results. For instance, if ECs driving style is very conservative, EC may prefer the lane-keeping and decelerating decision. However, if ECs driving style is very aggressive, it may accelerate and change lanes frequently. For other traffic scenarios, the similar conclusion can be drawn. Therefore, the decision-making result is an equilibrium of the game between

EC and ACs, which is related to all kinds of factors, i.e., motion states, driving behaviors, driving styles, etc.

FIGURE 5.1: Human-like lane-change decision making.

5.2.1.2 System Framework for Human-Like Decision Making

Figure 5.2 shows the human-like decision-making framework for AVs. In the modeling part, three kinds of different driving styles, i.e., aggressive, normal and conservative, are defined firstly. Then, a human-like driving model is proposed, which combines the driver model and the vehicle-road model. Considering different driving styles in the human-like driving model, human-like features are reflected in autonomous driving. Besides, different driving styles are considered in the decision-making cost function. In detail, different driving styles have different weights on the driving performances, including drive safety, ride comfort and travel efficiency. The decision-making cost function needs the real-time states and positions of surrounding obstacle vehicles. Based on the constructed decision-making cost function and multiple constraints, two game theoretic approaches, i.e., Nash equilibrium and Stackelberg game, are applied to the human-like decision making of AVs. Then, the decision-making results can be obtained, including the lane-change result, i.e., left lane change, lane keeping and right lane change, and the velocity planning result, i.e., accelerating, cruise and decelerating. Finally, the decision-making results will be outputted to the motion prediction and planning module designed with APF and MPC, which has been introduced in Chapter 4. The predicted motion states and positions are beneficial to the safe decision making of AVs.

In this chapter, we consider three kinds of driving styles defined in Chapter 2 for the human-like decision making of AVs, i.e., aggressive, normal and conservative [244, 225]. The aggressive driving style cares more about the travel efficiency. Therefore, this kind of driving style prefer higher speed. In contrast, the conservative driving style cares more the drive safety and ride comfort. Therefore, this kind of driving style prefer larger safe gap and smoother driving. The normal driving style is positioned between the above two styles,

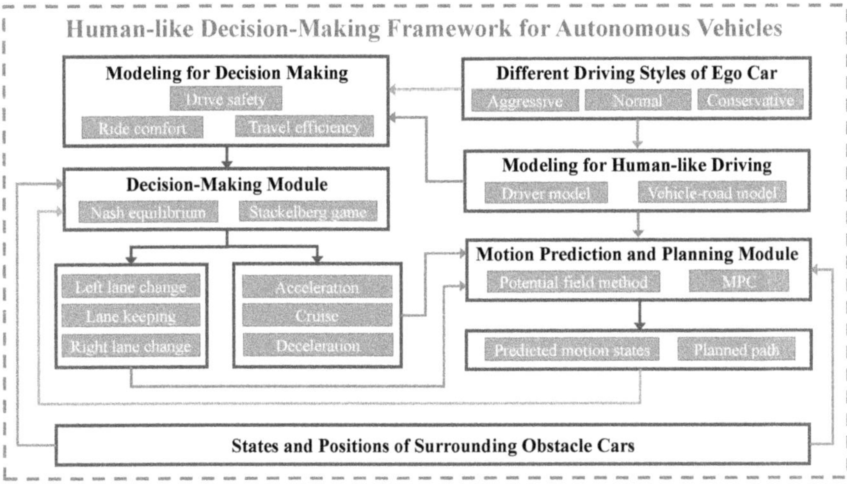

FIGURE 5.2: Human-like decision-making framework for AVs.

finding a good balance between different driving performances [147, 157]. The above three kinds of driving styles are reflected in the design process of the decision-making module and the trajectory planning module.

Besides, we use two noncooperative game theoretic approaches, i.e., the Nash equilibrium and Stackelberg game, to simulate the human-like interaction and decision making of AVs. The main difference between the two approaches is the decision-making logic. In the Nash equilibrium, all players are equal and make decisions at the same time. Each AV is an independent player and aims to minimize its individual decision-making cost function. However, in the Stackelberg game, there exists a leader player, and other players are follower. We consider EC as a leader player, and surrounding obstacle vehicles are assumed to be followers. The leader player makes the decision firstly, and the followers make the decision later. The leader can predict the decisions made by the followers and then make the optimal decision. The followers can only minimize the cost function after obtaining the decision-making results from the leader.

5.2.2 Human-like Decision Making Based on Noncooperative Game Theory

In this section, the decision-making cost function is constructed comprehensively considering safety, comfort and efficiency. Different driving styles are reflected in the decision-making cost function. Based on the constructed cost function, two game theoretic approaches, i.e., Nash equilibrium and Stackelberg game are used to address the human-like decision-making issue of AVs.

5.2.2.1 Cost Function for Lane-Change Decision Making

In the decision-making cost function, three driving performances are considered, including driving safety, ride comfort and travel efficiency. Taking EC as an instance, its decision-making cost function J^{EC} is constructed as follows.

$$J^{EC} = w_{ds}^{EC} J_{ds}^{EC} + w_{rc}^{EC} J_{rc}^{EC} + w_{te}^{EC} J_{te}^{EC} \tag{5.1}$$

where J_{ds}^{EC}, J_{rc}^{EC} and J_{te}^{EC} are the cost functions regarding driving safety, ride comfort and travel efficiency, respectively. w_{ds}^{EC}, w_{rc}^{EC} and w_{te}^{EC} denote the weighting coefficients.

The driving safety cost function of EC consists of the longitudinal driving safety cost and the lateral driving safety cost, which are affected by the motion of LC and AC, respectively. Then, the driving safety cost function of EC J_{ds}^{EC} is constructed as follows.

$$J_{ds}^{EC} = (\sigma^2 - 1)^2 J_{ds-log}^{EC} + \sigma^2 J_{ds-lat}^{EC} \tag{5.2}$$

where J_{ds-log}^{EC} and J_{ds-lat}^{EC} are the longitudinal driving safety cost and the lateral driving safety cost, respectively. σ denotes the lane-change behavior of EC, $\sigma \in \{-1, 0, 1\} \triangleq$\{left lane-change, lane keeping, right lane-change\}.

In detail, the longitudinal driving safety cost J_{ds-log}^{EC} is a function of the longitudinal gap and relative velocity between AC and LC, which is expressed as follows.

$$J_{ds-log}^{EC} = \kappa_{v-log}^{EC} \lambda_v^{EC} (\Delta v_{x,v}^{EC})^2 + \kappa_{s-log}^{EC} / [(\Delta s_{x,v}^{EC})^2 + \varepsilon] \tag{5.3}$$

$$\Delta v_{x,v}^{EC} = v_{x,v}^{LC} - v_{x,v}^{EC} \tag{5.4}$$

$$\Delta s_{x,v}^{EC} = [(X_v^{LC} - X_v^{EC})^2 + (Y_v^{LC} - Y_v^{EC})^2]^{1/2} - l_v \tag{5.5}$$

$$\lambda_v^{EC} = \begin{cases} 0, & \Delta v_{x,v}^{EC} \geq 0 \\ 1, & \Delta v_{x,v}^{EC} < 0 \end{cases} \tag{5.6}$$

where $v_{x,v}^{LC}$ and $v_{x,v}^{EC}$ are the longitudinal velocities of LC and EC, respectively. (X_v^{LC}, Y_v^{LC}) and (X_v^{EC}, Y_v^{EC}) denote the positions of LC and EC, respectively. κ_{v-log}^{EC} and κ_{s-log}^{EC} are the weighting coefficients. ε is a very small value to avoid zero denominator when conducting calculation. l_v is a safety coefficient considering the length of the car. v denotes the lane number, $v \in \{1, 2, 3\} \triangleq$\{left lane, middle lane, right lane\}.

Furthermore, the lateral driving safety cost J_{ds-lat}^{EC} is a function regarding the longitudinal gap and relative velocity between EC and AC, defined by

$$J_{ds-lat}^{EC} = \kappa_{v-lat}^{EC} \lambda_{v+\sigma}^{EC} (\Delta v_{x,v+\sigma}^{EC})^2 + \kappa_{s-lat}^{EC} / [(\Delta s_{x,v+\sigma}^{EC})^2 + \varepsilon] \tag{5.7}$$

$$\Delta v_{x,v+\sigma}^{EC} = v_{x,v}^{EC} - v_{x,v+\sigma}^{AC} \tag{5.8}$$

$$\Delta s_{x,v+\sigma}^{EC} = [(X_{v+\sigma}^{AC} - X_{v}^{EC})^2 + (Y_{v+\sigma}^{AC} - Y_{v}^{EC})^2]^{1/2} - l_v \tag{5.9}$$

$$\lambda_{v+\sigma}^{EC} = \begin{cases} 0, & \Delta v_{x,v+\sigma}^{EC} \geq 0 \\ 1, & \Delta v_{x,v+\sigma}^{EC} < 0 \end{cases} \tag{5.10}$$

where $v_{x,v+\sigma}^{AC}$ denotes the longitudinal velocity of AC, $(X_{v+\sigma}^{AC}, Y_{v+\sigma}^{AC})$ denotes the position of AC, κ_{v-lat}^{EC} and κ_{s-lat}^{EC} denote the weighting coefficients.

Besides, the ride comfort cost function J_{rc}^{EC} is expressed as a function regarding the longitudinal acceleration and lateral acceleration of EC, which is expressed as

$$J_{rc}^{EC} = \kappa_{a_x}^{EC}(a_{x,v}^{EC})^2 + \sigma^2 \kappa_{a_y}^{EC}(a_{y,v}^{EC})^2 \tag{5.11}$$

where $a_{x,v}^{EC}$ and $a_{y,v}^{EC}$ denote the longitudinal acceleration and lateral acceleration of EC. $\kappa_{a_x}^{EC}$ and $\kappa_{a_y}^{EC}$ denote the weighting coefficients.

Finally, the travel efficiency cost J_{te}^{EC} is related to the longitudinal velocity of EC, defined by

$$J_{te}^{EC} = (v_{x,v}^{EC} - \hat{v}_{x,v}^{EC})^2, \quad \hat{v}_{x,v}^{EC} = \min(v_{x,v}^{max}, v_{x,v}^{LC}) \tag{5.12}$$

where $v_{x,v}^{max}$ denotes the maximum travel velocity on the lane v.

Regarding the decision-making cost function of AC, it has the similar expression to EC. We will not introduce the work repeatedly. The difference is the decision-making vector. In this study, the lane-change behavior of AC is not considered. Only the longitudinal accelerating and decelerating behavior is considered. Therefore, the ride comfort cost function of AC is only related to the longitudinal acceleration.

To reflect the human-like characteristic in the decision-making issue of AVs, three kinds of driving styles, i.e., aggressive, normal and conservative, are considered in the weighting coefficient setting, i.e., w_{ds}^{EC}, w_{rc}^{EC} and w_{te}^{EC}. Based on the analysis results in [147, 157], the weighting coefficient setting regarding different driving styles is conducted in Table 5.1.

TABLE 5.1: Weighting Coefficients for Different Driving Styles

Driving Characteristic	Weighting Coefficients		
	w_{ds}^{EC}	w_{rc}^{EC}	w_{te}^{EC}
Aggressive	10%	10%	80%
Normal	50%	30%	20%
Conservative	70%	20%	10%

5.2.2.2 Noncooperative Decision Making Based on Nash Equilibrium

If there only exists one AC for EC, the lane-change decision-making issue of EC can be transformed into a 2-player game problem. Firstly, the Nash equilibrium game theoretic approach is used to address this issue, which is defined by

$$(a_{x,\upsilon}^{EC*}, \sigma^*) = \arg \max_{a_{x,\upsilon}^{EC}, \sigma} J^{EC}(a_{x,\upsilon}^{EC}, \sigma, a_{x,\upsilon+\sigma}^{AC}) \tag{5.13}$$

$$a_{x,\upsilon+\sigma}^{AC*} = \arg \max_{a_{x,\upsilon+\sigma}^{AC}} J^{AC}(a_{x,\upsilon}^{EC}, \sigma, a_{x,\upsilon+\sigma}^{AC}) \tag{5.14}$$

s.t. $\sigma \in \{-1,0,1\}$, $a_{x,\upsilon}^{EC} \in [a_{x,\upsilon}^{min}, a_{x,\upsilon}^{max}]$, $a_{x,\upsilon+\sigma}^{AC} \in [a_{x,\upsilon+\sigma}^{min}, a_{x,\upsilon+\sigma}^{max}]$, $v_{x,\upsilon}^{EC} \in [v_{x,\upsilon}^{min}, v_{x,\upsilon}^{max}]$, $v_{x,\upsilon+\sigma}^{AC} \in [v_{x,\upsilon+\sigma}^{min}, v_{x,\upsilon+\sigma}^{max}]$.

where $a_{x,\upsilon}^{EC*}$ and $a_{x,\upsilon+\sigma}^{AC*}$ denote the optimal longitudinal accelerations of EC and AC, σ^* denotes the optimal lane-change behavior of EC, $a_{x,i}^{min}$ and $a_{x,i}^{max}$ denote the minimum and maximum boundaries of the longitudinal acceleration, $v_{x,i}^{min}$ and $v_{x,i}^{max}$ denote the minimum and maximum control boundaries of the longitudinal velocity.

As Figure 5.1 shows, if there exist two ACs on the left and right lanes of EC, (5.13) and (5.14) can be further expanded to

$$(a_{x,\upsilon}^{EC*}, \sigma^*) = \arg \max_{(a_{x,\upsilon}^{EC}, \sigma) \in \Lambda} [J_1, J_2] \tag{5.15}$$

$$\Lambda = \{(a_{x,\upsilon,1}^{EC}, \sigma_1), (a_{x,\upsilon,1}^{EC}, \sigma_1)\} \tag{5.16}$$

$$J_1 = \arg \max_{a_{x,\upsilon,1}^{EC}, \sigma_1} J^{EC}(a_{x,\upsilon,1}^{EC}, \sigma_1, a_{x,\upsilon+\sigma_1}^{AC1}) \tag{5.17}$$

$$J_2 = \arg \max_{a_{x,\upsilon,2}^{EC}, \sigma_2} J^{EC}(a_{x,\upsilon,2}^{EC}, \sigma_2, a_{x,\upsilon+\sigma_2}^{AC2}) \tag{5.18}$$

$$a_{x,\upsilon+\sigma_1}^{AC1*} = \arg \max_{a_{x,\upsilon+\sigma_1}^{AC1}} J^{AC1}(a_{x,\upsilon,1}^{EC}, \sigma_1, a_{x,\upsilon+\sigma_1}^{AC1}) \tag{5.19}$$

$$a_{x,\upsilon+\sigma_2}^{AC2*} = \arg \max_{a_{x,\upsilon+\sigma_2}^{AC2}} J^{AC2}(a_{x,\upsilon,2}^{EC}, \sigma_2, a_{x,\upsilon+\sigma_2}^{AC2}) \tag{5.20}$$

s.t. $\sigma_1, \sigma_2 \in \{-1,0,1\}$, $a_{x,\upsilon,1}^{EC}, a_{x,\upsilon,2}^{EC} \in [a_{x,\upsilon}^{min}, a_{x,\upsilon}^{max}]$, $a_{x,\upsilon+\sigma_1}^{AC1} \in [a_{x,\upsilon+\sigma_1}^{min}, a_{x,\upsilon+\sigma_1}^{max}]$, $a_{x,\upsilon+\sigma_2}^{AC2} \in [a_{x,\upsilon+\sigma_2}^{min}, a_{x,\upsilon+\sigma_2}^{max}]$, $v_{x,\upsilon,1}^{EC}, v_{x,\upsilon,2}^{EC} \in [v_{x,\upsilon}^{min}, v_{x,\upsilon}^{max}]$, $v_{x,\upsilon+\sigma_1}^{AC1} \in [v_{x,\upsilon+\sigma_1}^{min}, v_{x,\upsilon+\sigma_1}^{max}]$, $v_{x,\upsilon+\sigma_2}^{AC2} \in [v_{x,\upsilon+\sigma_2}^{min}, v_{x,\upsilon+\sigma_2}^{max}]$.

5.2.2.3 Noncooperative Decision Making Based on Stackelberg Equilibrium

Besides the Nash equilibrium game theoretic approach, Stackelberg game theoretic approach is another noncooperative game approach [168, 169]. In the Nash equilibrium game theoretic approach, all players including EC and ACs are equal and independent. Namely, they make the decisions at the same time and independently. All players aims to minimize its individual decision-making cost function. However, in the Stackelberg game theoretic approach, there exists a leader and other players are followers. In the lane-change decision-making issue, EC is the leader player that makes the decision firstly. Other ACs are followers, and make the decisions later. For the case of one leader and one follower, the Stackelberg game theoretic approach for lane-change decision making is derived as follows.

$$(a_{x,v}^{EC*}, \sigma^*) = \arg \min_{a_{x,v}^{EC}, \sigma} \left(\max_{a_{x,v+\sigma}^{AC} \in \gamma^2(a_{x,v}^{EC}, \sigma)} J^{EC}(a_{x,v}^{EC}, \sigma, a_{x,v+\sigma}^{AC}) \right) \tag{5.21}$$

$$\gamma^2(a_{x,v}^{EC}, \sigma) = \{\zeta \in \Phi^2 : J^{AC}(a_{x,v}^{EC}, \sigma, \zeta) \leq J^{AC}(a_{x,v}^{EC}, \sigma, a_{x,v+\sigma}^{AC}), \forall a_{x,v+\sigma}^{AC} \in \Phi^2\} \tag{5.22}$$

s.t. $\sigma \in \{-1, 0, 1\}$, $a_{x,v}^{EC} \in [a_{x,v}^{min}, a_{x,v}^{max}]$, $a_{x,v+\sigma}^{AC} \in [a_{x,v+\sigma}^{min}, a_{x,v+\sigma}^{max}]$, $v_{x,v}^{EC} \in [v_{x,v}^{min}, v_{x,v}^{max}]$, $v_{x,v+\sigma}^{AC} \in [v_{x,v+\sigma}^{min}, v_{x,v+\sigma}^{max}]$.

For the case in Figure 5.1, i.e., one leader and two followers, (5.21) and (5.22) can be further expanded to

$$(a_{x,v}^{EC*}, \sigma^*) = \arg \max_{(a_{x,v}^{EC}, \sigma) \in \Lambda} [J^{EC}(a_{x,v,1}^{EC}, \sigma_1, a_{x,v+\sigma_1}^{AC1}), J^{EC}(a_{x,v,2}^{EC}, \sigma_2, a_{x,v+\sigma_2}^{AC2})] \tag{5.23}$$

$$\Lambda = \{(a_{x,v,1}^{EC}, \sigma_1), (a_{x,v,1}^{EC}, \sigma_1)\} \tag{5.24}$$

$$(a_{x,v,1}^{EC}, \sigma_1) = \arg \min_{a_{x,v,1}^{EC}, \sigma_1} \left(\max_{a_{x,v+\sigma_1}^{AC1} \in \gamma_1^2(a_{x,v,1}^{EC}, \sigma_1)} J^{EC}(a_{x,v,1}^{EC}, \sigma_1, a_{x,v+\sigma_1}^{AC1}) \right) \tag{5.25}$$

$$\begin{aligned} \gamma_1^2(a_{x,v,1}^{EC}, \sigma_1) = \{\zeta_1 \in \Phi_1^2 : J^{AC1}(a_{x,v,1}^{EC}, \sigma_1, \zeta_1) \leq \\ J^{AC1}(a_{x,v,1}^{EC}, \sigma_1, a_{x,v+\sigma_1}^{AC1}), \forall a_{x,v+\sigma_1}^{AC1} \in \Phi_1^2\} \end{aligned} \tag{5.26}$$

$$(a_{x,v,2}^{EC}, \sigma_2) = \arg \min_{a_{x,v,2}^{EC}, \sigma_2} \left(\max_{a_{x,v+\sigma_2}^{AC2} \in \gamma_2^2(a_{x,v,2}^{EC}, \sigma_2)} J^{EC}(a_{x,v,2}^{EC}, \sigma_2, a_{x,v+\sigma_2}^{AC2}) \right) \tag{5.27}$$

$$\begin{aligned} \gamma_2^2(a_{x,v,2}^{EC}, \sigma_2) = \{\zeta_2 \in \Phi_2^2 : J^{AC2}(a_{x,v,2}^{EC}, \sigma_2, \zeta_2) \leq \\ J^{AC2}(a_{x,v,2}^{EC}, \sigma_2, a_{x,v+\sigma_2}^{AC2}), \forall a_{x,v+\sigma_2}^{AC1} \in \Phi_2^2\} \end{aligned} \tag{5.28}$$

s.t. $\sigma_1, \sigma_2 \in \{-1, 0, 1\}$, $a_{x,v,1}^{EC}, a_{x,v,2}^{EC} \in [a_{x,v}^{min}, a_{x,v}^{max}]$, $a_{x,v+\sigma_1}^{AC1} \in [a_{x,v+\sigma_1}^{min}, a_{x,v+\sigma_1}^{max}]$, $a_{x,v+\sigma_2}^{AC2} \in [a_{x,v+\sigma_2}^{min}, a_{x,v+\sigma_2}^{max}]$, $v_{x,v,1}^{EC}, v_{x,v,2}^{EC} \in [v_{x,v}^{min}, v_{x,v}^{max}]$, $v_{x,v+\sigma_1}^{AC1} \in [v_{x,v+\sigma_1}^{min}, v_{x,v+\sigma_1}^{max}]$, $v_{x,v+\sigma_2}^{AC2} \in [v_{x,v+\sigma_2}^{min}, v_{x,v+\sigma_2}^{max}]$.

5.2.3 Testing Results and Performance Evaluation

To verify the two kinds of game theoretic human-like decision-making algorithm, two driving scenarios are designed and carried out based on the MATLAB-Simulink simulation platform.

5.2.3.1 Scenario A

Scenario A is illustrated in Figure 5.3, which is a typical merging case. Due to the impending ending of the current lane, EC must merge into the main lane. If EC makes the lane-change decision, it must interact with the AC on the main road. The driving behavior of AC have a significant effect on the decision-making result of EC. Besides, the driving style of EC is another factor to affect the decision-making results of EC. In this scenario, EC is defined with three kinds of driving styles, i.e., aggressive, normal and conservative. The initial velocities of EC and AC are set as 20 m/s and 15 m/s, and the initial gap between EC and AC is 2 m (EC is ahead). The test results are illustrated in Figures 5.4 to 5.6.

FIGURE 5.3: Testing Scenario A.

From Figure 5.4, we can find that different driving styles of EC result in various decision-making results. If EC is the aggressive mode, EC will speed up quickly and finish the lane-change process. The test results of velocities in Figure 5.5 can support the conclusion. It also indicates that the aggressive driving style prefers higher travel efficiency. However, if EC is the conservative mode, EC cares more about driving safety. Therefore, EC makes the decision slowing down and gives way for AC. From Figure 5.5, it can be found that EC has an obvious decelerating action after the seventh second. If EC is the normal mode, EC can finish the lane-change process, but its accelerating behavior is not faster than the aggressive mode. As a result, EC has to spend more time merging into the main lane. The detailed analysis of the test results is illustrated in Tables 5.2 and 5.3. In Table 5.2, t_c is the time of lane-change decision making, Δs_{t_c} is the gap between EC and AC at the time t_c. v_{x,t_c}^{EC} and v_{x,t_c}^{AC} are the velocities of EC and AC at the time t_c. We can see that the aggressive mode has the smallest t_c and largest v_{x,t_c}^{EC}, indicating that the aggressive driving style prefers higher travel efficiency. Besides, the conservative

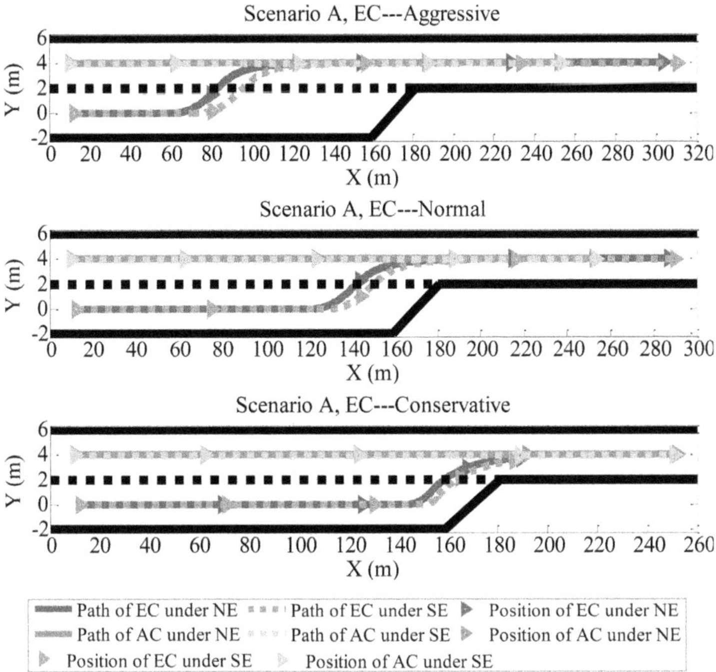

FIGURE 5.4: Decision-making results in Scenario A.

mode has the largest t_c and smallest v_{x,t_c}^{EC}, which means that the conservative driving style prefers higher driving safety. The test results of the normal mode are between the above two modes, indicating that the normal driving style aims to find a good balance between driving safety and travel efficiency. The cost values in Table 5.3 can support the conclusion again.

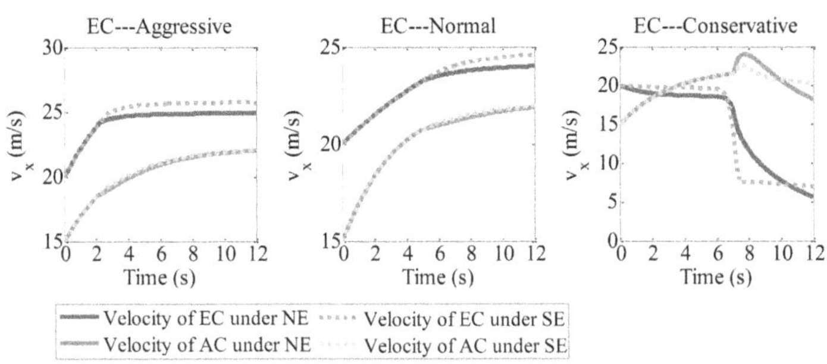

FIGURE 5.5: Testing results of velocities in Scenario A.

Although the test results of the two kinds of game theoretic decision-making algorithms, i.e., the Nash Equilibrium (NE) and Stackelberg

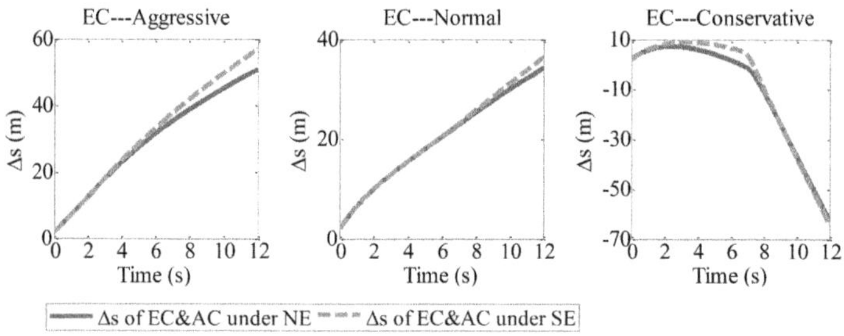

FIGURE 5.6: Testing results of the gaps in Scenario A.

TABLE 5.2: Testing Results of Decision Making in Scenario A

Testing Parameters	Aggressive		Normal		Conservative	
	NE	SE	NE	SE	NE	SE
t_c/s	2.24	2.72	5.04	5.36	7.00	7.18
$\Delta s_{t_c}/m$	13.77	16.46	18.06	18.86	-1.37	-1.99
$v_{x,t_c}^{EC}/(m/s)$	24.15	24.83	23.25	23.40	17.03	11.07
$v_{x,t_c}^{AC}/(m/s)$	18.66	19.17	20.76	20.91	21.49	21.52

TABLE 5.3: Cost Values on Driving Performances of EC in Scenario A

Cost Values ($\times 10^5$)	Aggressive		Normal		Conservative	
	NE	SE	NE	SE	NE	SE
Driving Safety	776	428	153	119	63	50
Ride Comfort	365	231	183	147	105	104
Travel Efficiency	114	79	275	215	394	329

Equilibrium (SE), are very similar, there still exist some differences. From Table 5.2, we can find that SE has larger t_c, Δs_{t_c} and v_{x,t_c}^{EC} than NE, which indicates that SE has better driving safety and higher travel efficiency than NE. That thanks to the predicted decision making of the follower player from the leader player. The cost values in Table 5.3 can support the conclusion. We can clearly find that with the SE approach, all the three kinds of driving performances, i.e., driving safety, ride comfort and travel efficiency, are improved. Taking the normal mode as an instance, the cost values of the driving safety, ride comfort and travel efficiency in SE are decreased by 22%, 20% and 22%, respectively.

5.2.3.2 Scenario B

In this scenario, a overtaking maneuver on a curved highway with three lanes is conducted, shown in Figure 5.7. EC and LC move on the middle lane. Besides, AC1 and AC2 move on the left and right lanes, respectively.

FIGURE 5.7: Testing Scenario B.

Due to the lower speed of LC, EC has to make a decision, lane keeping and slowing down to keep a safe distance between itself and LC, or lane-change and overtaking. If choosing lane-change, EC must interact with AC1 and AC2, and make the optimal decision, changing lanes to the left side or the right side? The driving behaviors of two ACs must be taken into consideration in the lane-change issue of EC. Moreover, the decision-making results are remarkably affected by the driving style of EC. Therefore, we consider three kinds of driving styles of EC, i.e., aggressive, normal and conservative. In this scenario, the initial velocities of LC, EC, AC1 and AC2 are set as 15 m/s, 20 m/s, 15 m/s and 13 m/s, respectively. The initial positions of LC, EC, AC1 and AC2 are set as (62, 0.95), (12, 0.04), (10, 4.03), and (15, -3.94), respectively. Two kinds of game theoretic decision-making algorithms are evaluated. The test results are illustrated in Figures 5.8 to 5.10.

TABLE 5.4: Testing Results of Decision Making in Scenario B

Testing Parameters	Aggressive		Normal		Conservative	
	NE	SE	NE	SE	NE	SE
t_c/s	2.98	3.44	3.90	4.20	–	–
$\Delta s^1_{t_c}/m$	28.66	33.32	25.53	26.20	–	–
$\Delta s^2_{t_c}/m$	26.22	31.21	22.55	24.06	–	–
$v^{EC}_{x,t_c}/(m/s)$	27.03	27.22	23.43	23.44	–	–
$v^{AC1}_{x,t_c}/(m/s)$	16.96	17.22	20.11	20.31	–	–
$v^{AC2}_{x,t_c}/(m/s)$	16.23	16.52	19.82	20.04	–	–

From Figure 5.8, we can find that different driving styles of EC lead to various decision-making results. If EC is the aggressive and normal mode, EC can finish the lane-change and overtaking process. However, if EC is the conservative mode, the decision-making result is lane-keeping. From Figure 5.9, we can see the velocity planning results. If EC is the aggressive mode, it has the a sudden accelerating behavior, with the shortest time to the largest

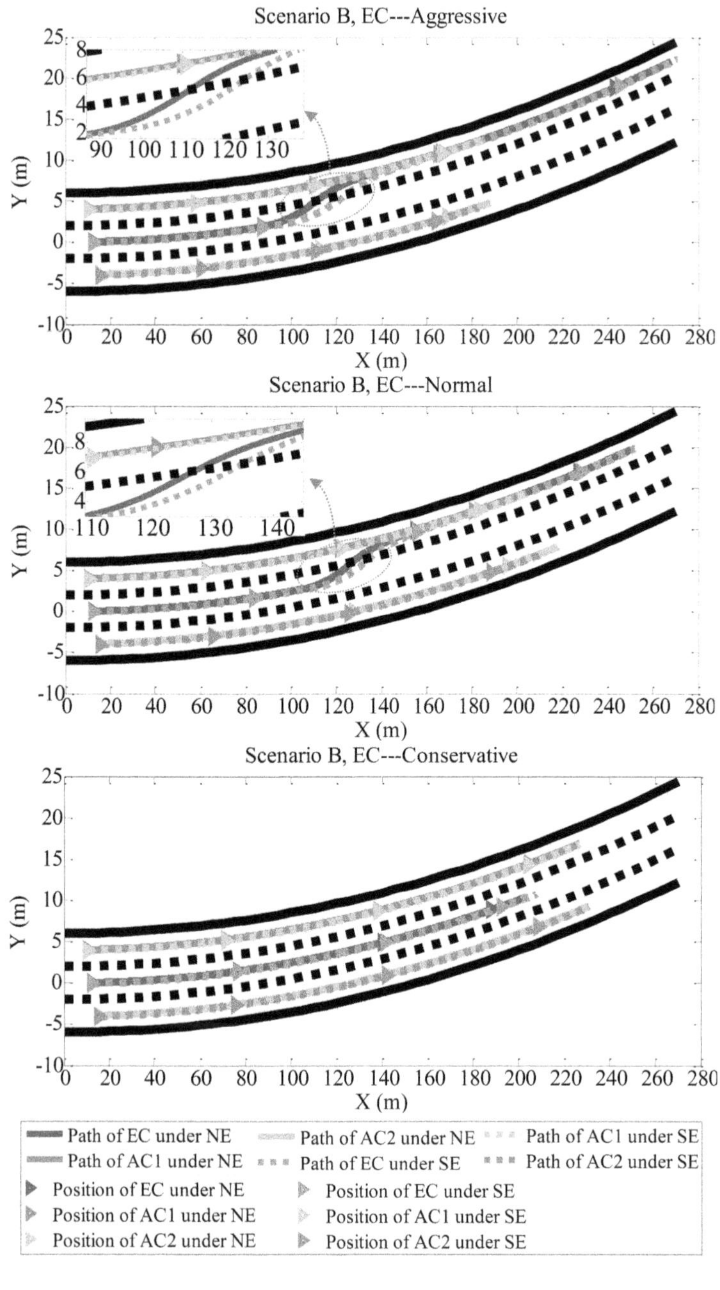

FIGURE 5.8: Decision-making results in Scenario B.

velocity, which indicates that the aggressive driving style prefers higher travel efficiency. However, if the driving style of EC is conservative, due to the lane-keeping decision, it has to decelerate after the fifth second to guarantee the

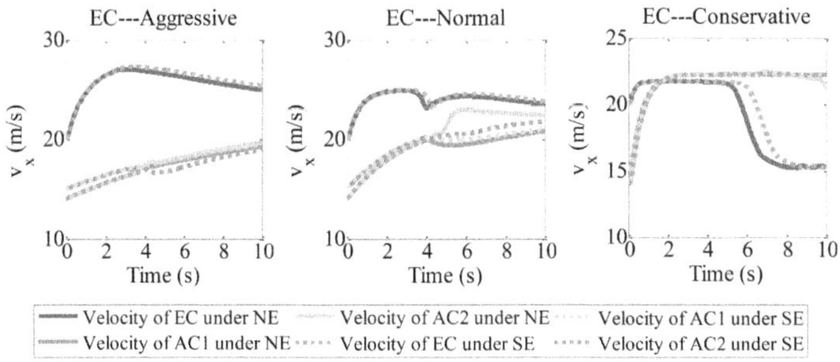

FIGURE 5.9: Testing results of velocities in Scenario B.

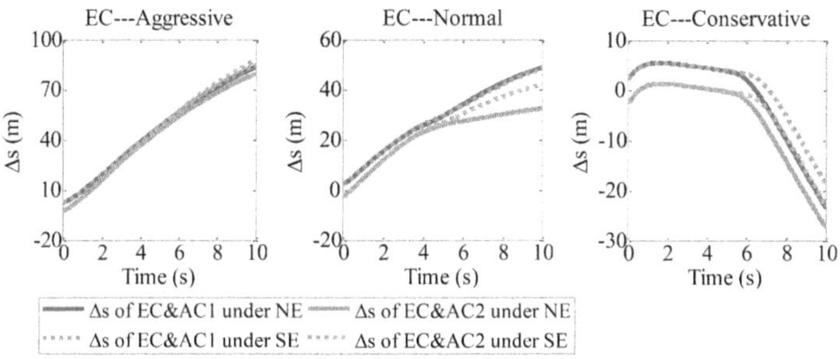

FIGURE 5.10: Testing results of the gaps in Scenario B.

TABLE 5.5: Cost Values on Driving Performances of EC in Scenario B

Cost Values ($\times 10^4$)	Aggressive		Normal		Conservative	
	NE	SE	NE	SE	NE	SE
Driving Safety	289	192	152	105	36	31
Ride Comfort	280	269	191	142	61	53
Travel Efficiency	119	107	279	205	371	344

driving safety. Figure 5.10 shows the gap between vehicles, the conservative driving style shows larger safe gap between vehicles than the aggressive driving style, indicating that the conservative driving style cares more about driving safety. The detailed decision-making results are listed in Table 5.4, where t_c is the time of lane-change decision making, $\Delta s^i_{t_c} (i = 1, 2)$ denotes the gap between EC and ACi at the time t_c. v^{EC}_{x,t_c} and $v^{AC1}_{x,t_c} (i = 1, 2)$ denote the velocities of EC and ACi at the time t_c. We can find that the aggressive driving style has smaller t_c and larger v^{EC}_{x,t_c} than the normal driving style, further indicating that the aggressive driving style prefers higher travel efficiency. Moreover, Table 5.5 shows the cost values of different driving styles on different driving performances. The conservative driving style has the smallest cost value on driving safety and ride comfort, which means that the conservative driving style pursues more on driving safety and ride comfort. However, the aggressive driving style has the smallest cost value on travel efficiency, further indicating that the aggressive driving style pursues more on travel efficiency.

Besides, there exist some differences on the decision-making results of two kinds of game theoretic approaches. The similar conclusion can be drawn that SE shows better performance than NE. Taking the normal driving style as an instance, compared with NE, the cost values of the driving safety, ride comfort and travel efficiency in SE are decreased by 31%, 34% and 27%, respectively. It indicates that the SE is beneficial to the performance improvement of driving safety, ride comfort and travel efficiency for EC.

5.2.3.3 Discussion of the Testing Results

Two different driving scenarios are conducted to verify the proposed human-like decision-making algorithms. In the test, different driving styles of EC are considered and two kinds of game theoretic approaches are evaluated. From the test results, we can find that different driving styles would result in different decision-making results of EC. The aggressive driving style would like to pursue higher travel efficiency, and the conservative style cares more about driving safety and ride comfort. Besides, the normal driving style tires to find a good balance between different driving performances. Moreover, both the two kinds of game theoretic approaches can make safe and reasonable decisions for vehicles. Compared with NE, due to the predicted decision making of the follower player from the leader player, SE shows better performance improvement of driving safety, ride comfort and travel efficiency for EC.

5.2.4 Summary

This study presents a human-like decision-making framework for AVs. Different driving styles are reflected in the design process of the decision-making cost function, including driving safety, ride comfort and travel efficiency. Based on the constructed cost function and multiple constraints, two kinds of game theoretic approaches, i.e., Nash equilibrium and Stackelberg game, are applied to the human-like decision making of AVs. Finally, two driving scenarios are designed and carried out to verify the human-like decision-making algorithms. The test results indicate that the proposed algorithms can make decisions for AVs like human drivers. The personalized decision making can be realized for different passengers.

5.3 Human-Like Decision Making of AVs at Unsignalized Roundabouts

In this study, with consideration of personalized driving styles, a game theoretic decision-making framework is designed to resolve the driving conflicts of AVs at an unsignalized roundabout. The Stackelberg game approach is advantageous with respect to the human-like interaction and decision making of AVs. The motion prediction of AVs is considered in the decision-making framework using MPC to enhance the effectiveness of the decision making. The interactions and decision making of AVs with different driving styles are studied, which provides personalized driving experience for passengers in terms of driving safety, ride comfort and travel efficiency. Finally, three test cases are designed and carried out to verify the feasibility, effectiveness, and real-time performance of the proposed human-like decision-making algorithm.

5.3.1 Problem Formulation and System Framework

5.3.1.1 Decision Making of AVs at Unsignalized Roundabouts

Although the driving conflict resolution of AVs at unsignalized roundabouts has been widely studied, most studies usually simplify the driving scenario as a single-lane roundabout. The driving conflict description is very simple, and the solution is to optimize the passing order and vehicle velocity. Actually, the driving conflicts at unsignalized roundabouts are very complex. The overtaking, lane change, merging, and other interaction behaviors among multi vehicles at unsignalized roundabouts should be taken into consideration. Unfortunately, these behaviors are usually neglected in most studies. To this end, this section presents a two-lane unsignalized roundabout scenario to study the interaction and conflict resolution of AVs.

Besides, most studies consider all AVs are the same driving characteristic. Therefore, in the driving conflict resolution at unsignalized roundabouts, all AVs usually show the similar decision-making behavior. The personalized driving demands are neglected. In this section, to realize human-like personalized driving, different driving styles are considered for AVs, reflecting different driving preference on safety, comfort and efficiency.

Figure 5.11 shows the two-lane unsignalized roundabout scenario. In this study, all vehicles are assumed to be AVs. The human-driven vehicles are not considered. From Figure 5.11, we can find that the unsignalized roundabout includes eight two-lane main roads and one two-lane round road, i.e., $M1$, $M2$, $M3$, $M4$, $\hat{M}1$, $\hat{M}2$, $\hat{M}3$, $\hat{M}4$ and RR. There exist four entrances and four exits, i.e., A_{in}, B_{in}, C_{in}, D_{in}, A_{out}, B_{out}, C_{out} and D_{out}. For decision-making algorithm design, the red car is defined as host vehicle (HV), and the blue cars are defined as the neighbor vehicles (NVs), which are the opponents

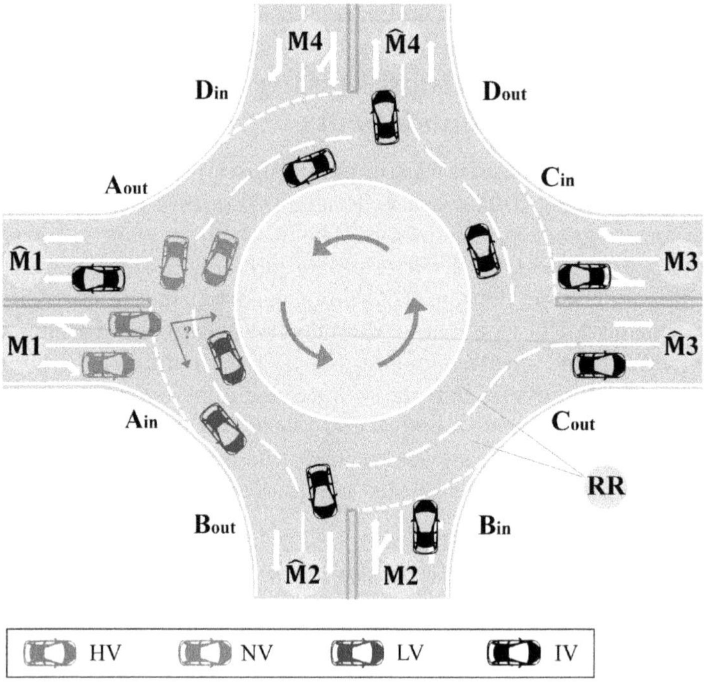

FIGURE 5.11: The scenario of the unsignalized roundabout with AVs.

to HV. The purple cars are defined as the lead vehicles (LVs), and the black cars are defined as the irrelevant vehicles (IVs). The interaction between HV and NV is double-sided. Namely, the decision-making behaviors of HV and NV are influenced by each other. However, the effect of LV on HV is unidirectional. Only the HV's decision-making behavior is affected by LV. Considering the large distance between IV and HV, the effect of IV on HV is neglected. It is worth mentioning that the name of HV, NV, LV and IV are relative. Each vehicle can be regarded as HV, then, the roles of NV, LV and IV are automatically generated according to the definition.

In general, the driving conflict resolution of AVs at the roundabout includes three steps, which is presented in Figure 5.12. The first step is the entering stage, shown in Figure 5.12(a). We can see that HV on the main road $M1$ prepares to merge into the round road RR from the entrance A_{in}. Firstly, it must interact with surrounding vehicles. The decision-making results are remarkably affected by the driving behaviors and driving styles of surrounding vehicles. HV has three choices, i.e., merging into the outside lane of the round road RR , merging into the inside lane of the round road RR, and lane keeping until stopping. If HV wants to merge into the outside lane of the round road, it has to fight for the right of way with NV1 and NV2. NV1 and NV2 have two choices, i.e., slowing down and yielding, or speeding up and fighting. The two

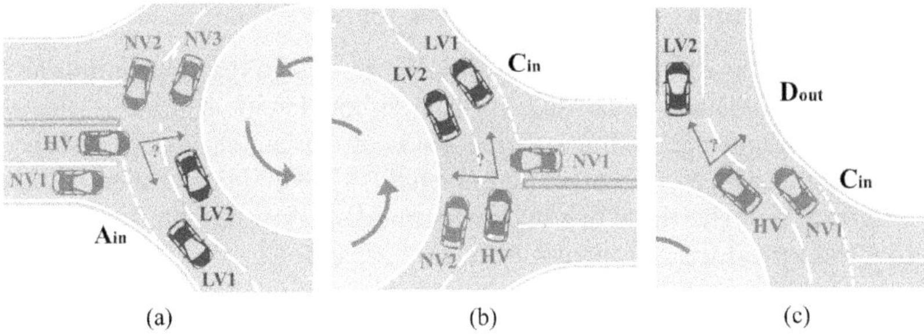

FIGURE 5.12: Decision-making process of AVs at the roundabout: (a) entering; (b) passing; (c) exiting.

decision-making results are remarkably affected by their driving styles. For instance, the conservative driving style prefers the former and the aggressive driving style prefers the later. Besides, if HV wants to merge into the inside lane of the round road, it has to interact with NV2 and NV3. Similarly, NV2 and NV3 have two choices, i.e., slowing down and yielding, or speeding up and fighting. The extreme condition is that HV slows down and gives ways for all NVs.

The middle process is the passing stage. In this state, HV must address the driving conflicts from the vehicles on the round road RR and the merging vehicles from the main road. Figure 5.12(b) presents an example. NV1 on the main road $M3$ starts to enter the round road RR from the entrance C_{in}. HV has three choices, i.e., slowing down and giving way to NV1, speeding up and fighting, changing lanes to the inside lane of the round road RR. The first two choices are associated with the driving behaviors of NV1 and HV. If HV chooses the third one, it must interact with NV2. Then, the driving conflict between HV and NV1 is converted into the driving conflict between HV and NV2.

The final step is to leave the roundabout. As Figure 5.12(c) shows, HV is getting ready to leave the roundabout from D_{out} and merge into the main road $\hat{M}4$. It must choose which lane of $\hat{M}4$ for entering. If HV makes the decision that entering the inside lane of $\hat{M}4$, it must follow LV2. If LV2 has a lower speed, HV has to slow down to keep a safe distance. Another choice for HV is lane-change. If so, it must interact with NV1 and fight for the right of way.

To sum up, the decision-making results of HV and all NVs at the unsignalized roundabout are affected by many factors, e.g., the relative velocity and distance, driving intention, and driving styles, which should be considered in the decision-making algorithm. This study aims to investigate the human-like interaction behavior and decision-making strategies of AVs at the complex unsignalized roundabout.

5.3.1.2 Decision-Making Framework for AVs

To address the driving conflicts of AVs at unsignalized roundabouts, a human-like decision-making framework is designed with the game theoretic approach. As Figure 5.13 shows, three kinds of driving styles are defined for AVs, i.e., aggressive, conservative and normal [147, 88]. The detailed definition and description of the three driving styles are displayed in Chapter 2. We will not give the introduction repeatedly in this section.

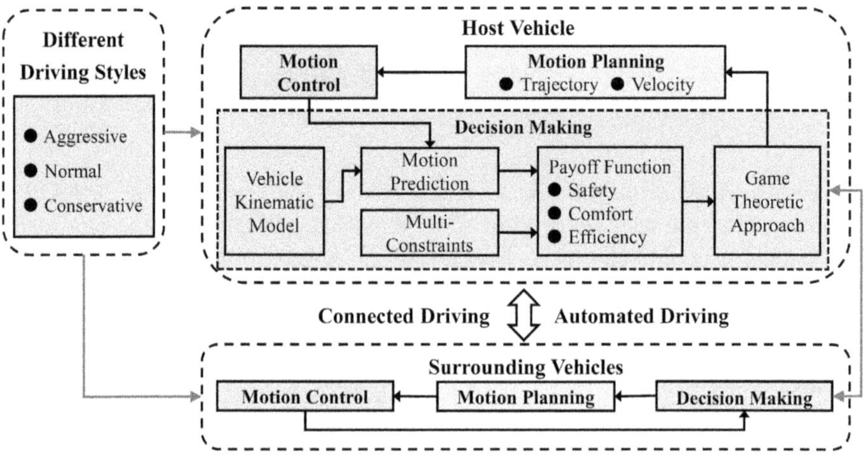

FIGURE 5.13: Decision-making framework for AVs.

Besides the driving style setting in Figure 5.13, a model-based motion prediction module is designed with the vehicle kinematic model in favor the performance improvement of the decision-making algorithm. In the decision-making payoff function design, different driving performances are considered including driving safety, ride comfort and travel efficiency. Based on the constructed payoff function and multiple constraints, the game theoretic approach is applied to the human-like interaction and decision making of AVs. Then, the formulated decision-making results are outputted to the motion planning and control modules. The motion planning and control algorithms have been presented in Chapter 4. This chapter aims to study the human-like decision-making algorithm.

5.3.2 Motion Prediction of AVs for Decision Making

For the model-based motion prediction algorithm design, a simplified vehicle kinematic model is adopted. Firstly, the vehicle kinematic model is expressed as a discrete model. With the given predictive horizon, the future motion states can be derived.

Firstly, the vehicle kinematic model (3.5) is expressed as a time-varying linear system.

$$\dot{x}(t) = A_t x(t) + B_t u(t) \tag{5.29}$$

where the time-varying coefficient matrices are derived by

$$A_t = \left.\frac{\partial f}{\partial x}\right|_{x_t, u_t}, \quad B_t = \left.\frac{\partial f}{\partial u}\right|_{x_t, u_t} \tag{5.30}$$

Then, (5.29) is expressed as a discrete model.

$$\begin{cases} x(k+1) = A_k x(k) + B_k u(k) \\ u(k) = u(k-1) + \Delta u(k) \end{cases} \tag{5.31}$$

where state vector $x(k) = [v_x(k), \varphi(k), X(k), Y(k)]^T$, coefficient matrices $A_k = e^{A_t \Delta T}$, $B_k = \int_0^{\Delta T} e^{A_t \tau} B_t d\tau$. Besides, ΔT denotes the sampling time, control vector $u(k) = [a_x(k), \delta_f(k)]^T$, control increment vector $\Delta u(k) = [\Delta a_x(k), \Delta \delta_f(k)]^T$.

Furthermore, a new state vector is proposed that combines the original state vector and the control vector.

$$\xi(k) = [x(k), u(k-1)]^T \tag{5.32}$$

After that, (5.31) is rewritten as

$$\begin{cases} \xi(k+1) = \hat{A}_k \xi(k) + \hat{B}_k \Delta u(k) \\ y(k) = \hat{C}_k \xi(k) \end{cases} \tag{5.33}$$

where $\hat{A}_k = \begin{bmatrix} A_k & B_k \\ 0_{2\times 4} & I_2 \end{bmatrix}$, $\hat{B}_k = \begin{bmatrix} B_k \\ I_2 \end{bmatrix}$ and $\hat{C}_k = \begin{bmatrix} I_4 & 0_{4\times 2} \end{bmatrix}$.

Defining the predictive horizon N_p and the control horizon N_c, $N_p > N_c$. At the time step k, if the state vector $\xi(k)$, the control vector $\Delta u(k)$ and coefficient matrices i.e., $\hat{A}_{p,k}$, $\hat{B}_{p,k}$ and $\hat{C}_{p,k}$ are known, the predicted state vectors are derived by

$$\begin{cases} \xi(p+1|k) = \hat{A}_{p,k} \xi(p|k) + \hat{B}_{p,k} \Delta u(p|k) \\ y(p|k) = \hat{C}_{p,k} \xi(p|k) \end{cases} \tag{5.34}$$

where $p = k, k+1, \cdots, k+N_p-1$.

Assuming that $\hat{A}_{p,k} = \hat{A}_k$, $\hat{B}_{p,k} = \hat{B}_k$ and $\hat{C}_{p,k} = \hat{C}_k$, the predicted motion states are expressed by

$$\begin{aligned}
\xi(k+1|k) &= \hat{A}_k \xi(k|k) + \hat{B}_k \Delta u(k|k) \\
\xi(k+2|k) &= \hat{A}_k^2 \xi(k|k) + \hat{A}_k \hat{B}_k \Delta u(k|k) + \hat{B}_k \Delta u(k+1|k) \\
&\vdots \\
\xi(k+N_c|k) &= \hat{A}_k^{N_c} \xi(k|k) + \hat{A}_k^{N_c-1} \hat{B}_k \Delta u(k|k) + \cdots \\
&\quad + \hat{B}_k \Delta u(k+N_c-1|k) \\
&\vdots \\
\xi(k+N_p|k) &= \hat{A}_k^{N_p} \xi(k|k) + \hat{A}_k^{N_p-1} \hat{B}_k \Delta u(k|k) + \cdots \\
&\quad + \hat{A}_k^{N_p-N_c} \hat{B}_k \Delta u(k+N_c-1|k)
\end{aligned} \tag{5.35}$$

The output vector sequence can be expressed by

$$\Upsilon(k) = [y^T(k+1|k), y^T(k+2|k), \cdots, y^T(k+N_p|k)]^T \tag{5.36}$$

According to (5.35) and (5.36), $\Upsilon(k)$ is derived by

$$\Upsilon(k) = \bar{C}\xi(k|k) + \bar{D}\Delta\mathbf{u}(k) \tag{5.37}$$

where $\Delta\mathbf{u}(k) = [\Delta u^T(k|k), \Delta u^T(k+1|k), \cdots, \Delta u^T(k+N_c-1|k)]^T$, $\bar{C} = [(\hat{C}_k\hat{A}_k)^T, (\hat{C}_k\hat{A}_k^2)^T, \cdots, (\hat{C}_k\hat{A}_k^{N_p})^T]^T$,

$$\bar{D} = \begin{bmatrix} \hat{C}_k\hat{B}_k & 0 & 0 & 0 \\ \vdots & \vdots & \vdots & \vdots \\ \hat{C}_k\hat{A}_k^{N_c-1}\hat{B}_k & \cdots & \hat{C}_k\hat{A}_k\hat{B}_k & \hat{C}_k\hat{B}_k \\ \vdots & \vdots & \vdots & \vdots \\ \hat{C}_k\hat{A}_k^{N_p-1}\hat{B}_k & \cdots & \hat{C}_k\hat{A}_k^{N_p-N_c+1}\hat{B}_k & \hat{C}_k\hat{A}_k^{N_p-N_c}\hat{B}_k \end{bmatrix}.$$

Finally, the model-based motion prediction of AVs is finished. The predicted motion states of AVs are used for the decision-making algorithm design, advancing the decision-making safety.

5.3.3 Algorithm Design of Decision Making Using the Game Theoretic Approach

In this section, the Stackelberg game approach is used to address the driving conflict of AVs at the unsignalized roundabout, and realize human-like decision making. Firstly, a decision-making payoff function is constructed comprehensively considering driving safety, ride comfort and travel efficiency. Combined with multiple constraints, the game theoretic decision making is transformed into an optimization issue.

5.3.3.1 Payoff Function of Decision Making Considering One Opponent

As mentioned above, the decision making of AVs at the unsignalized roundabout has three stages, i.e., entering, passing and exiting. Different stages have various decision-making behaviors. In the entering stage, the decision making of HV can be expressed as the merging process from the main road to the round road. Therefore, the driving conflicts come from the merging behavior. In the passing and exiting stage, the decision making of HV can be described as a lane-change issue. Thus, the driving conflicts are generated by the lane-change behavior. Furthermore, the merging behavior α is defined as follows, $\alpha \in \{-1, 0, 1\}$:= {*merging into the inside lane of round road, lane keeping, merging into the outside lane of round road*}. The lane-change behavior β is defined by $\beta \in \{-1, 0, 1\}$:= {*left lane change, lane keeping, right lane change*}. Considering that the driving conflict resolution of AVs at unsignalized roundabouts involves multiple vehicles and lanes, it brings a great challenge to design the decision-making payoff function. To simplify the design process, we firstly consider the lane-change issue that only contains two vehicles, i.e., HV and one NV. In the next section, the design approach will be generalized to multiple vehicles and lanes.

If AVi is regarded as HV, the payoff function for decision making can be expressed as follows, which comprehensively considers driving safety, ride comfort and travel efficiency.

$$P^i = k_s^i P_s^i + k_c^i P_c^i + k_e^i P_e^i \tag{5.38}$$

where P_s^i, P_c^i and P_e^i are the payoffs of driving safety, ride comfort and travel efficiency, respectively. k_s^i, k_c^i and k_e^i denote the weighting coefficients, associated with the driving style of AVi. According to the analysis results from human drivers' driving behaviors shown in Chapter 2, the weighting coefficients of the three different driving styles are depicted in Table 5.6.

Firstly, the payoff function of driving safety P_s^i consists of the longitudinal safety, lateral safety and lane-keeping safety, defined by

$$P_s^i = (1 - (\beta^i)^2)P_{s-log}^i + (\beta^i)^2 P_{s-lat}^i + (1 - (\beta^i)^2)P_{s-lk}^i \tag{5.39}$$

where P_{s-log}^i, P_{s-lat}^i and P_{s-lk}^i are the payoff functions regarding the longitudinal, lateral and lane keeping safety, respectively. We can find that if $\beta^i = 0$, i.e., lane-keeping, P_s^i is only related to the longitudinal safety and lane-keeping safety. Otherwise, when conducting lane-change, only lateral safety is considered.

TABLE 5.6: Weighting Coefficients of Different Driving Styles

Driving Characteristic	Weighting Coefficients		
	k_s^i	k_c^i	k_e^i
Aggressive	0.4	0.5	0.1
Normal	0.3	0.3	0.4
Conservative	0.1	0.4	0.5

Then, the payoff function of longitudinal safety P_{s-log}^i is related to the longitudinal gap and relative velocity between HV and its LV, which is expressed as

$$P_{s-log}^i = k_{v-log}^i[(\Delta v_{x,log}^i)^2 + \varepsilon]^{\eta_{log}^i} + k_{s-log}^i(\Delta s_{log}^i)^2 \tag{5.40}$$

$$\Delta v_{x,log}^i = v_x^{LV} - v_x^i \tag{5.41}$$

$$\Delta s_{log}^i = [(X^{LV} - X^i)^2 + (Y^{LV} - Y^{AVi})^2]^{1/2} - L_v \tag{5.42}$$

$$\eta_{log}^i = \text{sgn}(\Delta v_{x,log}^i) \tag{5.43}$$

where v_x^{LV} and v_x^i are the longitudinal velocities of LV and AVi, respectively. (X^{LV}, Y^{LV}) and (X^i, Y^i) denote the positions of LV and AVi, respectively. k_{v-log}^i and k_{s-log}^i denote the weighting coefficients. $\Delta v_{x,log}^i$ and Δs_{log}^i denote the relative velocity and longitudinal gap between LV and AVi. L_v denotes a safety coefficient considering the vehicle length. ε is a small positive value to avoid the zero denominator.

Besides, the payoff of lateral safety P_{s-lat}^i is a function of the relative distance and relative velocity between HV and NV, which is derived by

$$P_{s-lat}^i = k_{v-lat}^i[(\Delta v_{x,lat}^i)^2 + \varepsilon]^{\eta_{lat}^i} + k_{s-lat}^i(\Delta s_{lat}^i)^2 \tag{5.44}$$

$$\Delta v_{x,lat}^i = v_x^i - v_x^{NV} \tag{5.45}$$

$$\Delta s_{lat}^i = [(X^i - X^{NV})^2 + (Y^i - Y^{NV})^2]^{1/2} - L_v \tag{5.46}$$

$$\eta_{lat}^i = \mathrm{sgn}(\Delta v_{x,lat}^i) \tag{5.47}$$

where v_x^{NV} denotes the longitudinal velocity of NV. (X^{NV}, Y^{NV}) denotes the position of NV. k_{v-lat}^i and k_{s-lat}^i denote the weighting coefficients. $\Delta v_{x,lat}^i$ and Δs_{lat}^i denote the relative velocity and longitudinal gap between AVi and NV.

Moreover, the payoff of lane keeping safety P_{s-lk}^i is a function of the lateral distance error and the yaw angle error between the predicted position of AVi and the center line of the lane, expressed by

$$P_{s-lk}^i = k_{y-lk}^i/[(\Delta y^i)^2 + \varepsilon] + k_{\varphi-lk}^i/[(\Delta\varphi^i)^2 + \varepsilon] \tag{5.48}$$

where Δy^i and $\Delta\varphi^i$ denote the lateral distance error and yaw angle error, k_{y-lk}^i and $k_{\varphi-lk}^i$ denote the weighting coefficients.

The payoff function of ride comfort P_c^i is defined regarding the accelerations of AVi.

$$P_c^i = k_{a_x}^i/[(a_x^i)^2 + \varepsilon] + k_{a_y}^i/[(a_y^i)^2 + \varepsilon] \tag{5.49}$$

where a_x^i and a_y^i denote the longitudinal and lateral accelerations of AVi, $k_{a_x}^i$ and $k_{a_y}^i$ denote the weighting coefficients.

Additionally, the payoff function of travel efficiency P_e^i is associated with the longitudinal velocity of AVi, which is defined by

$$P_e^i = k_e^i/[(v_x^i - v_x^{\max})^2 + \varepsilon] \tag{5.50}$$

where $v_x^{\max}$ denotes the maximum travel velocity, and k_e^i denotes the weighting coefficient.

5.3.3.2 Payoff Function of Decision Making Considering Multiple Opponents

Based on the constructed decision-making payoff function, we can find that the driving safety payoff function involves multiple vehicles, especially for the longitudinal safety and lateral safety. The ride comfort payoff function and the travel efficiency payoff function are only related to motion states of HV. Namely, the ride comfort payoff function and the travel efficiency payoff function are not affected by the number of NVs. Therefore, in the decision-making payoff function that considers multiple opponents, we mainly focus on the derivation of the driving safety payoff function.

Considering the lane-change behavior and merging behavior of AVs at the unsignalized roundabout, the improved payoff function of driving safety is given by

$$P_s^i = \Gamma(P_{s-log}^i, P_{s-lat}^i, P_{s-lk}^i, \alpha^i, \beta^i) \tag{5.51}$$

where the function Γ is different at different decision-making stages.

At the entering stage, only the merging behavior is considered. Γ is derived by

$$\Gamma = (1-(\alpha^i)^2)P_{s-log}^i + (\alpha^i)^2 P_{s-lat}^i + (1-(\alpha^i)^2)P_{s-lk}^i \tag{5.52}$$

Besides, the payoff function of lateral safety is related to three NVs, which is written as

$$\begin{aligned} P_{s-lat}^i = {} & 0.25(\alpha^i+1)^2 P_{s-lat}^{i\&NV1} + P_{s-lat}^{i\&NV2} \\ & +0.25(\alpha^i-1)^2 P_{s-lat}^{i\&NV3} \end{aligned} \tag{5.53}$$

where $P_{s-lat}^{i\&NV1}$, $P_{s-lat}^{i\&NV2}$ and $P_{s-lat}^{i\&NV3}$ denote the payoffs of lateral safety regarding NV1, NV2 and NV3, respectively.

At the passing and exiting stage, only the lane-change behavior is considered in Γ, i.e.,

$$\Gamma = (1-(\beta^i)^2)P_{s-log}^i + P_{s-lat}^i + (1-(\beta^i)^2)P_{s-lk}^i \tag{5.54}$$

From Figure 5.12, we can see that the payoff of lateral safety is not only related to NV1, but also affected by the merging behavior of NV2. Furthermore, P_{s-lat}^i is derived by

$$P_{s-lat}^i = (\beta^i)^2 P_{s-lat}^{i\&NV1} + (\alpha^{NV2})^2 P_{s-lat}^{i\&NV2} \tag{5.55}$$

where α^{NV2} denotes the merging behavior of NV2.

5.3.3.3 Constraints of Decision Making

In the decision-making issue of AVs at unsignalized roundabouts, AVs are limited within the control boundaries. Namely, AVs should follow some constraints, including the safety constraints, control constraints, etc.

Firstly, the safety constraints are proposed.

$$|\Delta s^i| \le \Delta s^{\max}, |\Delta y^i| \le \Delta y^{\max}, |\Delta \varphi^i| \le \Delta \varphi^{\max} \tag{5.56}$$

Besides, the constraints for ride comfort are expressed by

$$|a_x^i| \le a_x^{\max}, |a_y^i| \le a_y^{\max} \tag{5.57}$$

Moreover, the velocity constraint is given by

$$|v_x^i| \le v_x^{\max} \tag{5.58}$$

Additionally, the control constraint of Δa_x^i is defined as

$$|\Delta a_x^i| \leq \Delta a_x^{\max} \tag{5.59}$$

The control constraints of $\Delta \delta_f^i$ and δ_f^i are given by

$$|\Delta \delta_f^i| \leq \Delta \delta_f^{\max}, |\delta_f^i| \leq \delta_f^{\max} \tag{5.60}$$

Finally, the above constraints for AVi is rewritten in a compact form.

$$\Xi^i(\Delta s^i, \Delta y^i, \Delta \varphi^i, a_x^i, a_y^i, v_x^i, \Delta a_x^i, \Delta \delta_f^i, \delta_f^i) \leq 0 \tag{5.61}$$

The parametric values of the constraints are depicted in Table 5.7.

TABLE 5.7: Constraint Boundaries for Decision Making

Parameter	Value	Parameter	Value
$\Delta s^{\max}$/ (m)	0.8	$v_x^{\max}$/ (m/s)	30
$\Delta y^{\max}$/ (m)	0.2	$\Delta a_x^{\max}$/ (m/s^2)	0.1
$\Delta \varphi^{\max}$/ (deg)	2	$\Delta \delta_f^{\max}$/ (deg)	0.3
$a_x^{\max}$/ (m/s^2)	8	$\delta_f^{\max}$/ (deg)	30
$a_y^{\max}$/ (m/s^2)	5	–	–

5.3.3.4 Decision Making with the Game Theory and MPC Optimization

As Figure 5.14 shows, the Stackelberg game approach is used to resolve the driving conflicts of AVs at the unsignalized roundabout. In the Stackelberg game, there exits a leader and some followers. Both the leader and followers aim to maximize their decision-making payoffs. Differing from the normal noncooperative game, the leader makes the decision firstly, and the followers make the decisions later. The leader is endowed the power of predicting follower's strategy given its own[169]. Therefore, the followers can only maximize their payoffs under the leader's decision. The leader has a first-mover advantage.

The decision-making vector of AVi is defined as $\hat{u}^i = [\Delta a_x^i, \Delta \delta_f^i, \alpha^i, \beta^i]^T$. Based on the constructed payoff function, the Stackelberg game theoretic approach for decision making is derived by.

$$\begin{aligned} \hat{u}^{HV*} = \arg \max_{\hat{u}^{HV} \in U^{HV}} (\min_{\Lambda} P^{HV}(\hat{u}^{HV}, \hat{u}^{NV1}, \\ \hat{u}^{NV2}, \hat{u}^{NV3})) \end{aligned} \tag{5.62}$$

$$\begin{aligned} \Lambda \triangleq \{\hat{u}^{NV1} \in U^{NV1*}, \hat{u}^{NV2} \in U^{NV2*}, \\ \hat{u}^{NV3} \in U^{NV3*}\} \end{aligned} \tag{5.63}$$

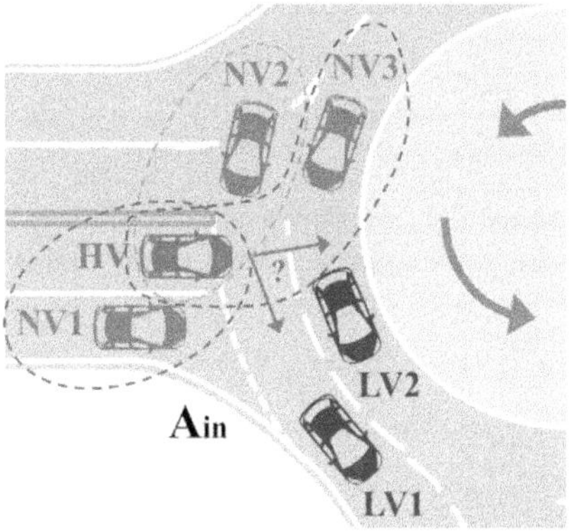

FIGURE 5.14: Decision making of AVs at the roundabout with the Stackelberg game approach.

$$\begin{aligned} U^{NV1*}(\hat{u}^{HV}) \triangleq \{ & \hat{u}^{NV1*} \in U^{NV1} : P^{NV1}(\hat{u}^{HV}, \\ & \hat{u}^{NV1*}) \geq P^{NV1}(\hat{u}^{HV}, \hat{u}^{NV1}), \forall \hat{u}^{NV1} \in U^{NV1} \} \end{aligned} \tag{5.64}$$

$$\begin{aligned} U^{NV2*}(\hat{u}^{HV}) \triangleq \{ & \hat{u}^{NV2*} \in U^{NV2} : P^{NV1}(\hat{u}^{HV}, \\ & \hat{u}^{NV2*}) \geq P^{NV2}(\hat{u}^{HV}, \hat{u}^{NV2}), \forall \hat{u}^{NV2} \in U^{NV2} \} \end{aligned} \tag{5.65}$$

$$\begin{aligned} U^{NV3*}(\hat{u}^{HV}) \triangleq \{ & \hat{u}^{NV3*} \in U^{NV3} : P^{NV3}(\hat{u}^{HV}, \\ & \hat{u}^{NV3*}) \geq P^{NV3}(\hat{u}^{HV}, \hat{u}^{NV3}), \forall \hat{u}^{NV3} \in U^{NV3} \} \end{aligned} \tag{5.66}$$

s.t. $\Xi^{HV} \leq 0$, $\Xi^{NV1} \leq 0$, $\Xi^{NV2} \leq 0$, $\Xi^{NV3} \leq 0$.
where $\hat{u}^{HV*}, \hat{u}^{NV1*}, \hat{u}^{NV2*}$ and $\hat{u}^{NV3*}$ are the optimal decision-making results of HV, NV1, NV2 and NV3, respectively.

Combined with the motion prediction of AVs, the MPC approach is applied to the decision-making optimization.

At the time step k, the predicted payoff function sequence for AVi is derived as follows.

$$P^i(k+1|k), P^i(k+2|k), \cdots, P^i(k+N_p|k) \tag{5.67}$$

To realize prediction-based decision making with MPC optimization, the following cost function is constructed.

$$J^i = 1/(P^i + \varepsilon) \tag{5.68}$$

where the coefficient $\varepsilon \to 0, \varepsilon > 0$.

The decision-making sequence for AVi is expressed by

$$\hat{u}^i(k|k), \hat{u}^i(k+1|k), \cdots, \hat{u}^i(k+N_c-1|k) \tag{5.69}$$

where $\hat{u}^i(q|k) = [\Delta a_x^i(q|k), \Delta\delta_f^i(q|k), \alpha^i(q|k), \beta^i(q|k)]^T$, $q = k, k+1, \cdots, k+N_c-1$.

Then, the performance function of AVi for decision making is derived by

$$\Pi^i = \sum_{p=k+1}^{k+N_p} ||J^i(p|k)||_Q^2 + \sum_{q=k}^{k+N_c-1} ||\hat{u}^i(q|k)||_R^2 \tag{5.70}$$

where Q and R denote the weighting matrices.

Finally, the Stackelberg game approach for decision making is realized with MPC optimization.

$$\begin{aligned} &(\hat{\mathbf{u}}^{HV*}, \hat{\mathbf{u}}^{NV1*}, \hat{\mathbf{u}}^{NV2*}, \hat{\mathbf{u}}^{NV3*}) = \arg\min \Pi^{HV} \\ &\text{s.t.} \quad \min \Pi^{NV1}, \min \Pi^{NV2}, \min \Pi^{NV3}, \\ &\quad \Xi^{HV} \leq 0, \Xi^{NV1} \leq 0, \Xi^{NV2} \leq 0, \Xi^{NV3} \leq 0. \end{aligned} \tag{5.71}$$

where $\hat{\mathbf{u}}^{HV*}$, $\hat{\mathbf{u}}^{NV1*}$, $\hat{\mathbf{u}}^{NV2*}$ and $\hat{\mathbf{u}}^{NV3*}$ are the optimal decision-making sequences of HV, NV1, NV2 and NV3.

Although only one HV and three NVs are discussed in the above optimization, the proposed decision-making algorithm can be generalized to more vehicles via the increase of the decision-making vector and payoff function.

5.3.4 Testing Results and Analysis

This section aims to evaluate the feasibility and effectiveness of the human-like decision-making algorithm. Three test cases are designed for the unsignalized roundabout driving scenario. All the tests are carried out with the MATLAB/Simulink simulation platform. To realize human-like decision making, AVs are given different driving styles in the test.

5.3.4.1 Testing Case 1

This case mainly focuses on resolving the driving conflicts at the entering stage. HV on the inside lane of the main road $M1$ is ready to enter the round road RR from the entrance A_{in} and exit from B_{out} to the main road $\hat{M}2$. In this driving scenario, HV must interact with three NVs, i.e., NV1 on the outside lane of the main road $M1$, NV2 on the outside lane of the round road RR and NV3 on the inside lane of the round road RR, and finally make the optimal decision.

To realize personalized driving and human-like decision making, AVs are given different driving styles in three driving scenarios. In Scenario A, the driving styles of HV, NV1, NV2 and NV3 are set as normal, conservative, normal and normal, respectively. In Scenario B, the driving styles of the four AVs are set as normal, aggressive, normal and normal, respectively. In Scenario C, the driving styles of the four AVs are set as aggressive, aggressive, normal and normal, respectively. In Case 1, the initial positions of HV, NV1, NV2, NV3, LV1 and LV2 are set as (-25, -2.45), (-28, -6.08), (-16, 10.25), (-10, 11.18), (-16, -10.25) and (-14, -5.38), respectively. In addition, the initial velocities of HV, NV1, NV2, NV3, LV1 and LV2 are set as 5.5 m/s, 4 m/s, 5 m/s, 4 m/s, 8 m/s, 8 m/s, respectively. Finally, the test results are presented in Figures 5.15 to 5.20.

From Figure 5.15 we can find that different driving styles of AVs result in various decision-making results. In Scenario A, HV decided to speed up and merge into the

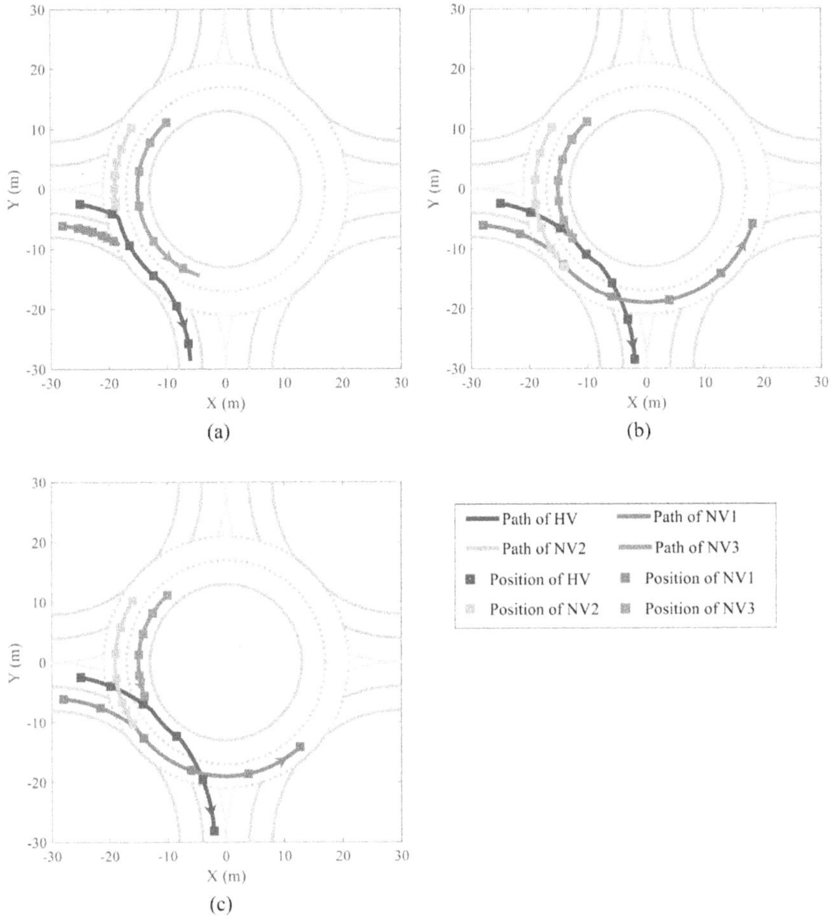

FIGURE 5.15: Decision-making results of AVs in Case 1: (a) Scenario A; (b) Scenario B; (c) Scenario C.

outside lane of the round road *RR*. NV1 and NV2 slowed down and gave ways for HV. The merging behavior of HV has no effect on NV3. The main reason that HV made this decision is the conservative driving style of NV1. NV1 decelerated to advance the driving safety. In Scenario B, the driving style of NV1 is aggressive. NV1 is not willing to give ways for HV. As a result, NV1 has a sudden acceleration shown in Figure 5.16, indicating that NV1 pursues higher travel efficiency. Then, HV decided to merge into the inside lane of the round road *RR*. Before merging into the roundabout, HV has a deceleration behavior to keep a safe distance between itself and NV2. Finally, NV3 yielded and slowed down to give ways for HV. In Scenario C, both HV and NV1 are aggressive. The competition is more intense. Finally, HV merged into the inside lane of the round road *RR*. Both NV2 and NV3 yielded and

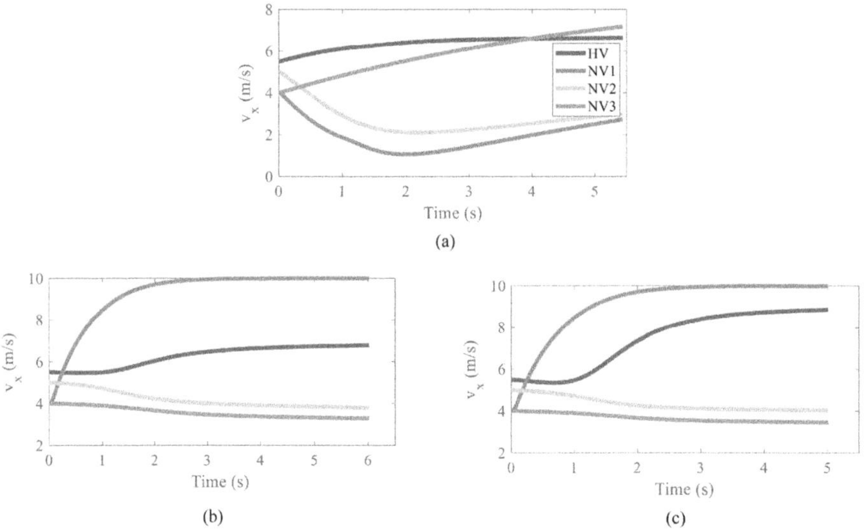

FIGURE 5.16: Velocities of AVs in Case 1: (a) Scenario A; (b) Scenario B; (c) Scenario C.

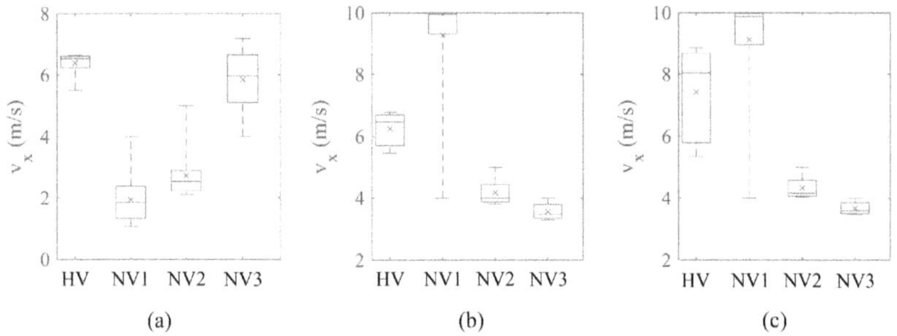

FIGURE 5.17: Box plots of velocities in Case 1: (a) Scenario A; (b) Scenario B; (c) Scenario C.

decelerated. We can find from Figure 5.16 that both HV and NV1 have larger travel velocity. In general, the aggressive driving style pursue high travel efficiency. The velocity distribution in Figure 5.17 can support the conclusion as well.

Besides, the distributions of longitudinal and lateral accelerations in this case are illustrated in Figures 5.18 and 5.19, respectively. From the test results, we can find that the aggressive driving style shows larger maximum acceleration and distribution. However, the conservative driving style shows smaller acceleration distribution, which indicates that the conservative driving style cares more about ride comfort than the aggressive driving style. Moreover, the relative distances between HV and

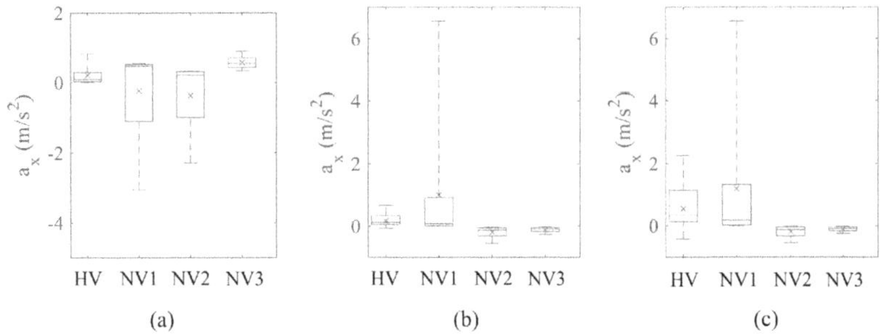

FIGURE 5.18: Box plots of longitudinal accelerations in Case 1: (a) Scenario A; (b) Scenario B; (c) Scenario C.

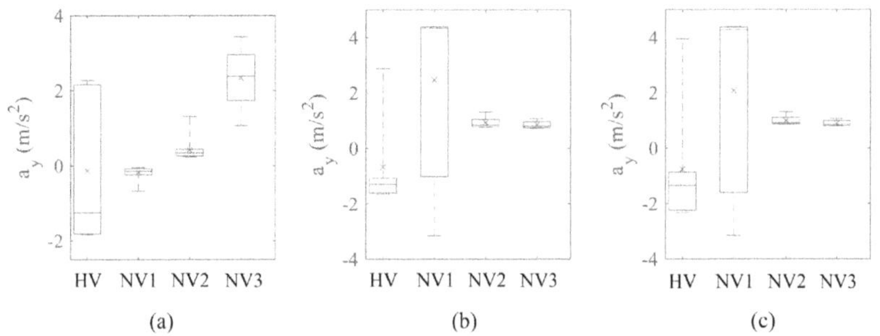

FIGURE 5.19: Box plots of lateral accelerations in Case 1: (a) Scenario A; (b) Scenario B; (c) Scenario C.

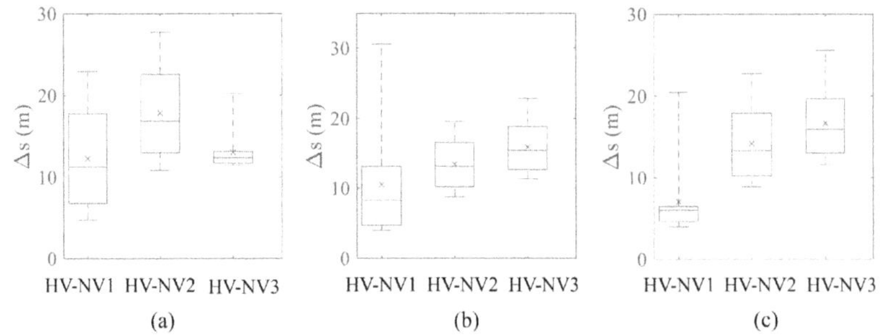

FIGURE 5.20: Box plots of relative distances between HV and other NVs in Case 1: (a) Scenario A; (b) Scenario B; (c) Scenario C.

other NVs are shown in Figure 5.20. We can find that the driving safety is worsen for the aggressive driving style. Especially for two aggressive drivers, the relative distance is too small. However, the conservative driving style brings larger relative distance in favor the improvement of driving safety.

5.3.4.2 Testing Case 2

In this case, the passing and exiting stages of the unsignalized roundabout are discussed. HV on the outside lane of the round road RR is ready to move out from D_{out} to the main road $\hat{M}4$. In this process, HV has to resolve the merging conflict from NV1 on the main road $M3$. HV has three choices, i.e., slowing down and giving ways to NV1, speeding up and fighting for the right of way, or changing lanes to the inside lane of the round road RR. If HV chooses the last one, it must interact with NV2. The merging conflict from NV1 is transformed into a lane-change conflict.

In this case, to simulate the human-like driving and decision making, different driving styles are set for AVs. In Scenario A, the driving styles of HV, NV1 and NV2 are set as conservative, normal and normal, respectively. In Scenario B, the driving styles of the three AVs are all normal. In Scenario C, the driving styles of the three AVs are set as aggressive, normal and normal, respectively. Besides, the initial positions of HV, NV1, NV2, LV1 and LV2 are set as (15, -11.66), (25, 2.45), (8, -12.68), (15, 11.66) and (13, 7.48), respectively. The initial velocities of HV, NV1, NV2, LV1 and LV2 are set as 5.5 m/s, 5 m/s, 4 m/s, 8 m/s, 5 m/s, respectively. Finally, the test results under different scenarios are illustrated in Figures 5.21 to 5.26.

From the decision-making results in Figure 5.21, we can find that the driving style has significant effect on the decision making of AVs. In Scenario A, due to the conservative driving style, HV cares more about driving safety. Therefore, HV made the decision slowing down and giving ways for NV1, and finally exited to the outside lane of the main road $\hat{M}4$. From Figure 5.22, we can see a sudden deceleration of HV. Scenario B shows the different decision-making result of HV. HV made the lane-change decision to give ways for NV1, and finally exited to the inside lane of the main road $\hat{M}4$. As a result, NV2 yielded and slowed down. We can see an obvious deceleration of NV2 in Figure 5.22. In Scenario, the driving style of HV is aggressive. Therefore, HV pursues higher travel efficiency. We can see that HV conducted a double-lane change to deal with the driving conflict, and finally exited to the outside lane of the main road $\hat{M}4$. NV1 and NV2 have to slow down and give ways for HV.

Besides, Figures 5.24 and 5.25 show the distribution of the longitudinal and lateral accelerations, respectively. We can find that the aggressive driving style has larger acceleration distribution, and the conservative driving styles shows smaller acceleration distribution. It can be concluded that the conservative driving style prefers better ride comfort, and the aggressive driving style pursues higher travel efficiency. Figure 5.26 shows the test results of relative distances between HV and NVs. It can be found that the aggressive driving style tends to shorten the safe distance and the conservative driving style would like to enlarge the relative distance to advance driving safety.

5.3.4.3 Testing Case 3

This case considers five AVs at the unsignalized roundabout. All the three stages are included in this case, and the travel efficiency of the entire traffic system is evaluated. At the initial time, HV on the outside lane of the main road $M1$ is ready to enter the round road RR from the entrance A_{in} and exit from A_{out} to the main road $\hat{M}1$, simulating a U-turn scenario. Besides HV, four NVs are considered at the unsignalized roundabout. NV1 on the inner lane of the round road RR starts to exit

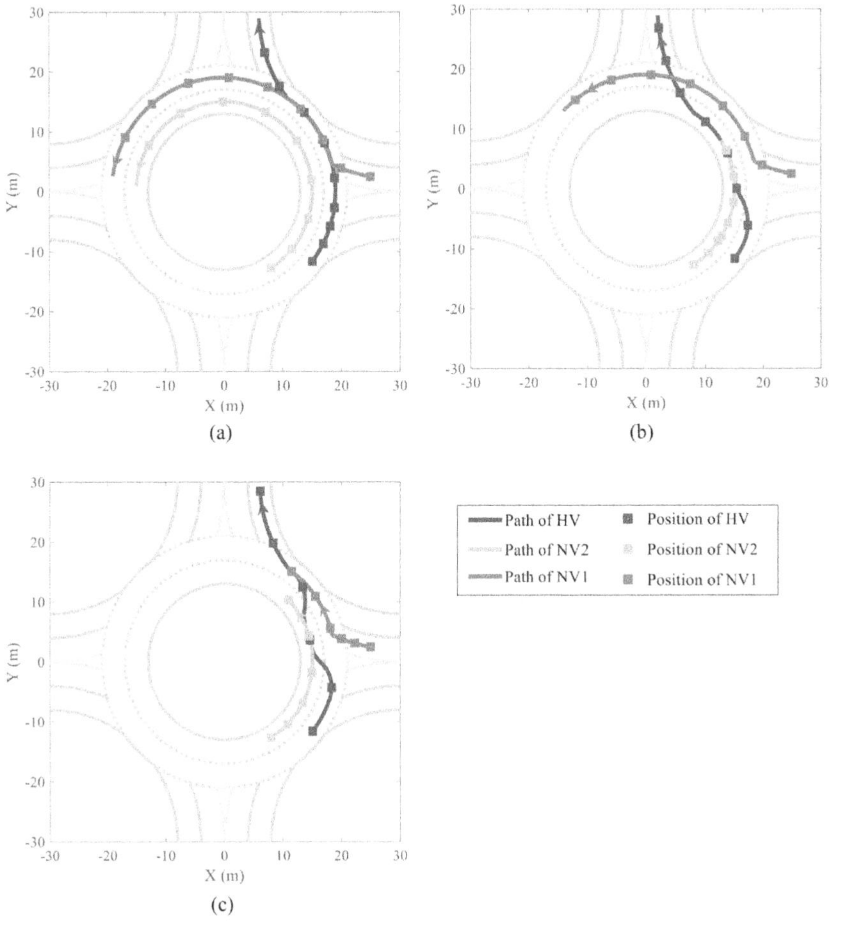

FIGURE 5.21: Decision-making results of AVs in Case 2: (a) Scenario A; (b) Scenario B; (c) Scenario C.

from C_{out} to the main road $\hat{M}3$. NV2 on the external lane of the main road $M2$ is ready to enter the round road RR and exit from C_{out} to the main road $\hat{M}3$. NV3 on the inner lane of the main road $M3$ starts to enter the round road RR and exit from A_{out} to the main road $\hat{M}1$. NV4 on the outer lane of the main road $M4$ wants to enter the round road RR from the entrance D_{in} and exit from D_{out} to the main road $\hat{M}4$.

To study the effect of personalized driving on the travel efficiency of the entire traffic system, different driving styles are considered for AVs. The driving style of HV is set as aggressive and the driving styles of four NVs are set as normal. The initial positions of HV, NV1, NV2, NV3 and NV4 are set as (-19, -8.68), (-9, -12), (6, -35), (50, 2) and (-6, 88), respectively. The initial velocities of HV, NV1, NV2,

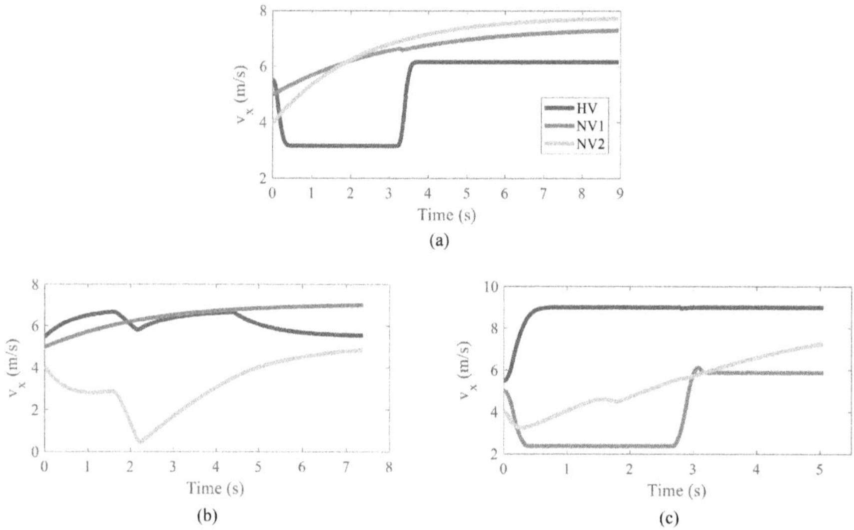

FIGURE 5.22: Velocities of AVs in Case 2: (a) Scenario A; (b) Scenario B; (c) Scenario C.

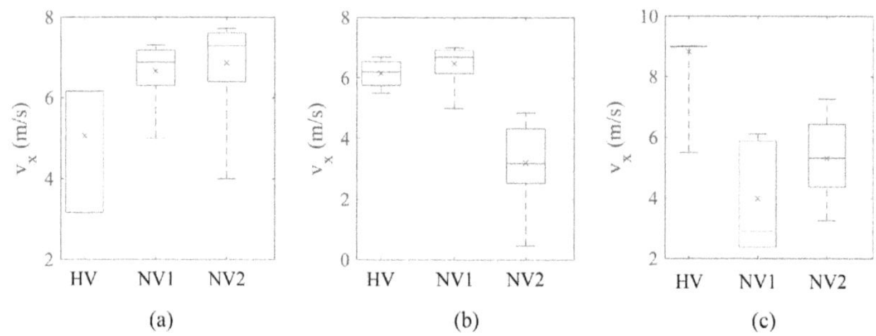

FIGURE 5.23: Box plots of velocities in Case 2: (a) Scenario A; (b) Scenario B; (c) Scenario C.

NV3 and NV4 are set as 5.5 m/s, 5 m/s, 5 m/s, 4 m/s, 4 m/s, respectively. Finally, the test results are illustrated in Figures 5.27 to 5.29.

From Figure 5.27, we can find that there exist three driving conflict zones for HV. The first driving conflict comes from the merging behavior of NV2. HV made the lane-change decision to resolve this conflict. In this stage, we can see from Figure 5.28 that HV has an obvious deceleration to guarantee the safe distance between itself and NV1. The second driving conflict is the following conflict between HV and NV3. Due to the lower travel velocity of NV3, HV gave up following and changed lanes to the external lane of the round road. The third driving conflict is the merging

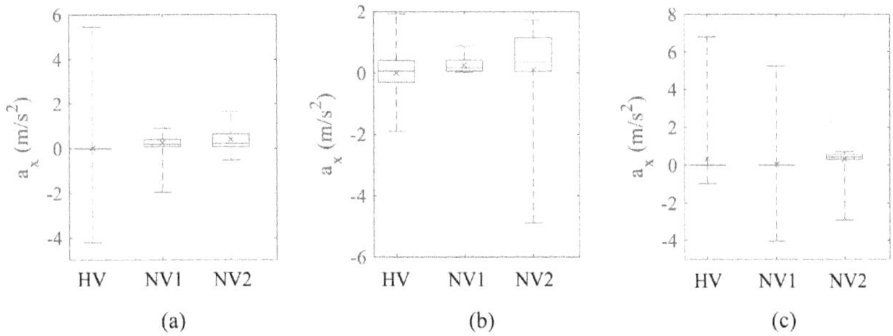

FIGURE 5.24: Box plots of longitudinal accelerations in Case 2: (a) Scenario A; (b) Scenario B; (c) Scenario C.

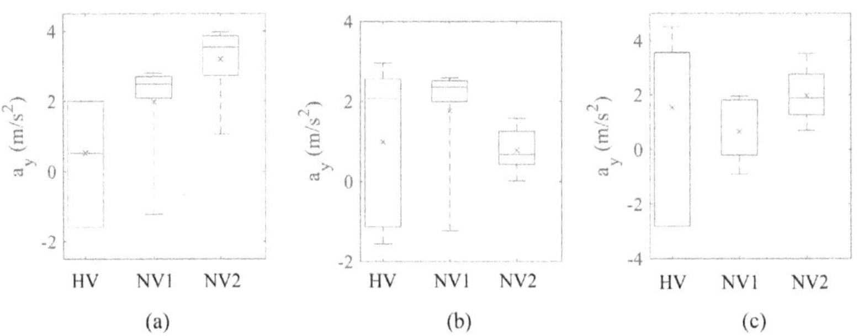

FIGURE 5.25: Box plots of lateral accelerations in Case 2: (a) Scenario A; (b) Scenario B; (c) Scenario C.

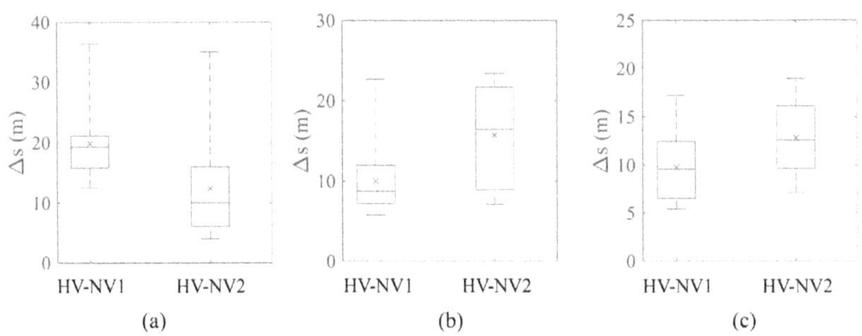

FIGURE 5.26: Box plots of relative distances between HV and other NVs in Case 2: (a) Scenario A; (b) Scenario B; (c) Scenario C.

conflict from NV4. HV made the decision slowing down to give ways for NV4. We can see from Figure 5.28 that the aggressive driving style prefers frequency acceleration and deceleration behaviors. Table 5.8 shows the detailed analysis of the vehicle velocity and system velocity. We can find that HV has the largest velocity among

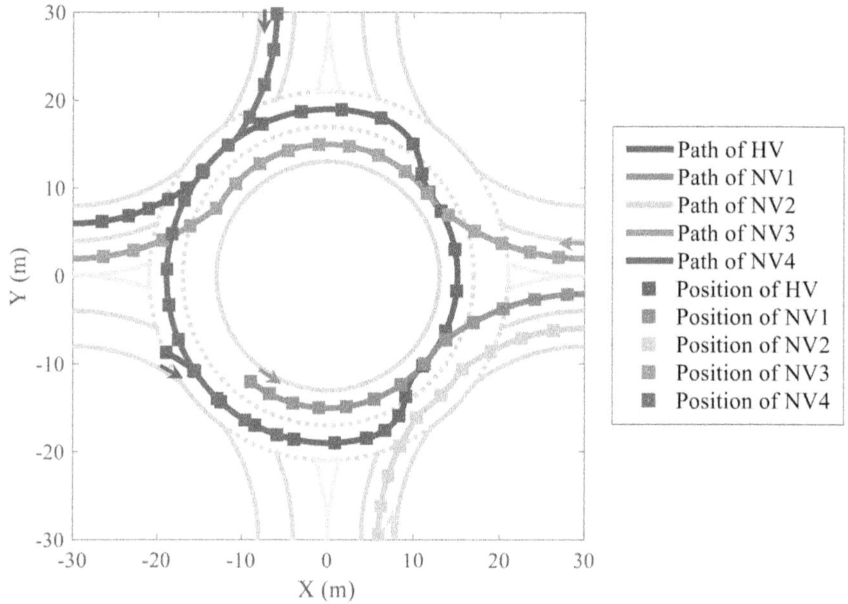

FIGURE 5.27: Decision-making results of AVs in Case 3.

all vehicles, which indicates that the aggressive driving style pursues higher travel efficiency. Moreover, we can find that the due to the aggressive driving of HV, the travel efficiency of NV3 is decreased. As a result, the travel efficiency of the entire traffic system is decreased.

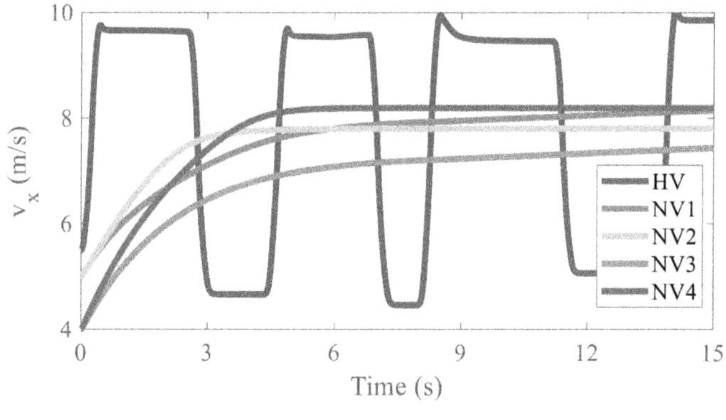

FIGURE 5.28: Velocities of AVs in Case 3.

TABLE 5.8: Travel Efficiency Analysis in Case 3

Testing Results	HV	NV1	NV2	NV3	NV4
Velocity Max / (m/s)	10.06	8.14	7.80	7.43	8.19
Velocity RMS / (m/s)	8.10	7.59	7.57	6.87	7.76
System Velocity RMS / (m/s)			7.59		

Besides, the computational time of the game-theoretic decision-making algorithm is illustrated in Figure 5.29. The mean value of the computational time for each iteration is 0.050s.

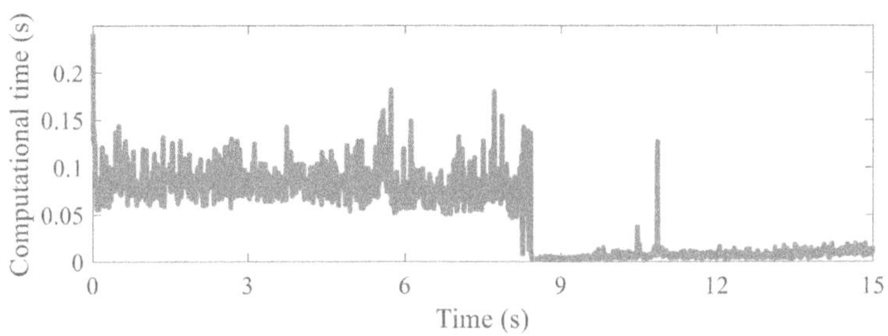

FIGURE 5.29: Computational time of the proposed algorithm.

Based on the analysis of the test results, we can conclude that the proposed game-theoretic decision-making algorithm can help AVs resolve the driving conflicts at the unsignalized roundabout. Besides, personalized driving and human-like decision making can be realized. In general, the aggressive driving style pursues higher travel efficiency, and the conservative driving style prefers better driving safety and ride comfort. It is worth mentioning that the proposed game-theoretic decision-making algorithm can be generalized to more complex driving scenarios via the adjustment of the decision-making vector and payoff function.

5.3.5 Summary

In this study, a human-like decision-making framework is applied to the conflict resolution of AVs at the unsignalized roundabout. Firstly, different driving styles are proposed for AVs to realize personalized driving. A model-based motion-prediction algorithm is designed with a simplified vehicle kinematic model, in favor of the performance improvement of the decision-making algorithm. In the decision-making payoff function, different driving performances are considered including driving safety, ride comfort and travel efficiency. Based on the payoff function and multiple constraints, the Stackelberg game theoretic approach is applied to the decision making of AVs. Finally, the human-like decision-making framework is verified with three test cases. The test results indicate that the game theoretic approach can help AVs make safe decisions at the unsignalized roundabout. Besides, personalized driving and human-like decision making can be realized as well.

5.4 Conclusion

This chapter aims to construct the human-like decision-making framework for AVs. Different driving scenarios are tested, including lane-change on highways and decision making at the unsignalized roundabout. To realize human-like decision making, different driving styles are set for AVs, including conservative, normal, and aggressive, yielding different decision-making cost functions regarding driving safety, ride comfort and travel efficiency. In the human-like lane-change decision-making issue, two noncooperative game theoretic approaches are utilized, i.e., Nash equilibrium and Stackelberg game. Based on the decision-making cost functions and multiple constraints, the interaction and decision-making process of AVs is realized with the noncooperative game theoretic approaches. It is found that both methods are able to generate reasonable and proper decisions for AVs, while the Stackelberg game is advantageous over the other one with respect to the resultant performance of the AV during decision making. Moreover, different driving styles lead to different decision results. The aggressive driving style prefers frequent lane-change to pursue higher travel efficiency, while the conservative driving style cares more about driving safety.

Besides the lane-change scenario, the decision-making issue of AVs at the unsignalized roundabout is studied as well. Different driving styles are set for AVs to simulate human-like driving. Based on the decision-making payoff function and multiple constraints, the Stackelberg game theoretic approach is applied to the driving conflict resolution of AVs at the unsignalized roundabout. The test results show that the noncooperative game theoretic approach can make safe decisions for AVs at the unsignalized roundabout. Additionally, the human-like decision making can be realized.

Chapter 6

Decision Making for CAVs with Cooperative Game Theoretic Method

DOI: 10.1201/9781003287087-6

6.1 Background

As the number of cars increases, traffic congestion is becoming a distressing issue. Especially during peak hours, the traffic efficiency is worsened remarkably, leading to the increase of energy consumption and the risk of collisions [91, 290, 195]. Taking the highway scenario as an example, vehicle merging is a typical condition resulting in traffic congestion [85]. In existing traffic rules, the vehicles on the main lane have higher right of way. Therefore, the vehicles on the on-ramp lane should slow down and give ways to the vehicles on the main lane [133]. However, if the vehicle on the on-ramp lane is aggressive, it will speed up suddenly and forcibly change lanes. As a result, the vehicles on the main road have to decelerate to give ways. The aggressive driving behavior of the vehicle on the on-ramp lane will worsen the travel efficiency of the main lane, even causes traffic jams and accidents. On the other hand, if the vehicle on the on-ramp lane is conservative, to guarantee the driving safety, the vehicle on the on-ramp lane may look for a safe enough distance before changing lanes, which will lead to more time for merging and the traffic congestion on the on-ramp lane. Therefore, the merging timing of vehicles on the on-ramp is associated with many factors, e.g., vehicle velocity, relative distance, and driving style, which has a significant effect on the driving safety and travel efficiency.

Besides the highway scenario, the urban scenario is more likely to cause traffic congestion, e.g., the unsignalized roundabout. Autonomous and connected driving technology is effective to address the above issue, improving driving safety and travel efficiency [17]. In a connected driving environment, vehicles can share their motion velocities, positions, driving behaviors and even driving intentions with the vehicle to vehicle (V2V) technology [14]. As a result, CAVs can easily make the safe decisions with the collaborative decision making [231]. The driving conflicts in the complex traffic scenarios can be resolved, and the traffic efficiency can be improved remarkably [31, 137].

In existing studies of the collaborative decision making of CAVs, CAVs are usually regarded as the same agent [235]. Namely, all CAVs do not have the personalized driving behaviors. They must follow the order of the centralized controller at the complex conflict zone [157]. The collaborative decision making of CAVs is not sovereign and spontaneous. In this chapter, the cooperative game theoretic approaches are applied to the collaborative decision making of CAVs at the merging conflict zone and the unsignalized roundabout. Both the individual interest and the system interest will be considered.

6.2 Cooperative Lane-Change and Merging of CAVs on Highways

This section aims to resolve the driving conflicts of CAVs at the multi-lane merging zone. Based on the coalitional game approach, a collaborative decision-making framework is proposed. To advance the decision-making safety, a model-based motion prediction module is designed with a simplified single-track vehicle model. Comprehensively considering all kinds of driving performances, e.g., driving safety, ride comfort and travel efficiency, the decision-making cost function is constructed. To realize personalized driving in the collaborative decision making, different driving styles are considered for CAVs, including aggressive, normal and conservative defined in Chapter 2. Different driving styles show various weighting allocations on safety, comfort and efficiency. In this section, four typical coalition models are proposed for the collaborative decision making of CAVs at the multi-lane merging zone. Based on the constructed decision-making cost function and multiple constraints, the coalitional game theoretic approach is used to resolve the merging conflicts of CAVs at the multi-lane merging zone. To verify the proposed decision-making algorithm, two test cases are designed and carried out. The test results indicate that the proposed coalitional game theoretic approach can address the driving conflicts at the multi-lane merging zone, making safe decisions for CAVs. Besides, it can realize collaborative decision making for CAVs to improve the traffic efficiency. Most importantly, the personalized driving demands are considered in the collaborative decision-making process.

6.2.1 Problem Formulation and System Framework

6.2.1.1 Problem Formulation

As mentioned in Chapter 1, the centralized control system has been widely studied to resolve the merging conflicts of CAVs at the multi-lane merging zone. Although it can plan the safe merging sequence and velocities for CAVs, all CAVs must submit their control rights to the centralized control system. As a result, CAVs cannot realize sovereign decision making and the personalized driving is sacrificed, which is a damage to the individual interest. For instance, the autonomous ambulance cares more about travel efficiency. It should have higher passing priority at the multi-lane merging zone. Considering that future CAVs must be designed in personalized and human-like ways, we proposed the collaborative decision-making framework for CAVs with the coalitional game theoretic approach. According to different driving objectives and driving styles, CAVs can form different coalition types for the collaborative decision making.

In most studies regarding the merging conflict resolution of CAVs at the merging zone, only two lanes are considered, one main lane and one on-ramp lane. Actually, the main road usually has more lanes. Therefore, the lane-change behavior of CAVs on the main road is neglected in the two-lane merging scenario. Most studies only considers the merging sequence optimization and velocity planning of CAVs, which is unreal for the actual merging scenario. To this end, a multi-lane merging scenario is studied in this study. As Figure 6.1 shows, the merging scenario on the highway includes two main lanes and one on-ramp lane. For instance, V1 is ready to change lanes and merge into the main road, which will yield a merging conflict between itself and V2 on the main lane. Therefore, before changing lanes, V1 must interact with V2 firstly. V2 has three choices, i.e., slowing down and giving ways for V1, speeding up and fighting the right of way, and changing lanes to the left side and giving ways for V1. If V2 makes the third decision, the merging conflict between V1 and V2 will be transformed into the lane-change conflict between V2 and V3. V3 can make two kinds of decisions, slowing down and yielding, or speeding up and

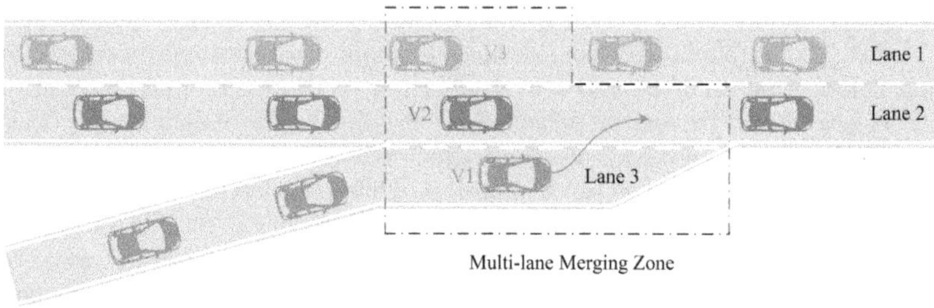

FIGURE 6.1: Decision making of CAVs at the multi-lane merging zone.

fighting the right of way. All the decisions of V1, V2 and V3 are affected by many factors, including the relative velocity and distance between vehicles, and driving styles. For instance, if V2 is the conservative driving style, it prefers better driving safety. As a result, V2 would like to slow down and give ways for V1. However, if V2 is the aggressive driving style, it pursues higher travel efficiency. Therefore, V2 prefers speeding up and fighting for the right of way. Besides the personalized driving, CAVs can cooperate with each other to conduct the collaborative decision making.

6.2.1.2 Cooperative Decision-Making Framework for CAVs

To address the above issue, a cooperative decision-making framework is designed for CAVs in Figure 6.2. To realize human-like decision decision making, different driving styles are considered in the decision-making algorithm design, including aggressive, normal and conservative, reflecting different preferences on driving safety, ride comfort and travel efficiency [157, 78]. The definition and analysis of the three kinds of driving styles are presented in Chapter 2.

In the decision-making module, a motion prediction sub-module is included, which is designed with a simplified vehicle model, further advancing the decision-making safety. In the decision-making cost function design, three critical driving performances are considered, i.e., driving safety, ride comfort, and travel efficiency. Based on the constructed decision-making cost functions of host CAVs and surrounding CAVs, combing multiple constraints, the coalitional game theoretic approach is applied to the collaborative decision making of CAVs. In the coalitional game theoretic approach, CAVs at the conflict zone will form different coalitions. All CAVs in the same coalition try to minimize the coalitional cost. Finally, the coalition will output the decision-making results of each CAV, e.g., the merging behavior, acceleration and steering angle. The path planning and motion control algorithms have been introduced in Chapter 4. We will not give the repeated description. All vehicle are assumed to be CAVs in this section, and the information can be shared with the V2V technology.

6.2.2 Motion Prediction of CAVs

To advance the decision-making safety of CAVs, a model-based motion prediction algorithm is derived based on the single-track vehicle model described in Chapter 3.

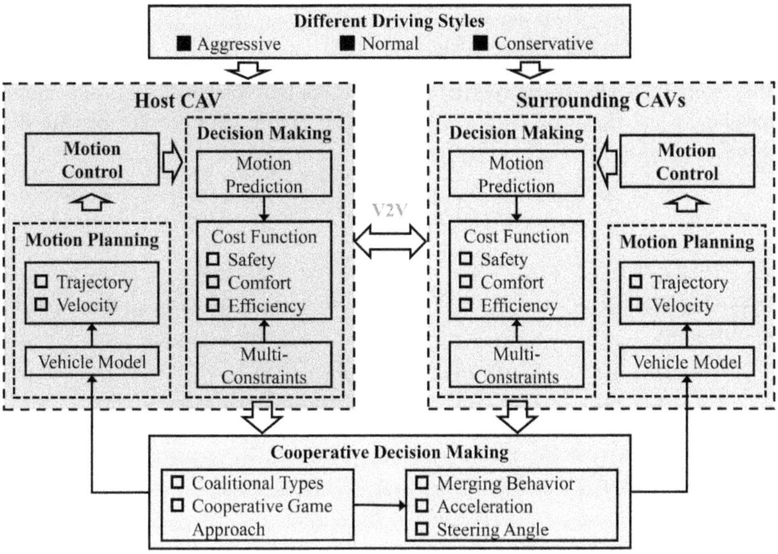

FIGURE 6.2: The proposed decision-making framework for CAVs.

Firstly, the single-track vehicle model is rewritten as a time-varying linear system as follows.

$$\dot{x}(t) = A_t x(t) + B_t u(t) \tag{6.1}$$

where the time-varying coefficient matrices are defined by

$$A_t = \left.\frac{\partial \Gamma}{\partial x}\right|_{x_t, u_t}, \quad B_t = \left.\frac{\partial \Gamma}{\partial u}\right|_{x_t, u_t} \tag{6.2}$$

Then, (6.1) is expressed in the discrete form.

$$\begin{cases} x(k+1) = A_k x(k) + B_k u(k) \\ u(k) = u(k-1) + \Delta u(k) \end{cases} \tag{6.3}$$

where sate vector $x(k) = [v_x(k), v_y(k), r(k), \varphi(k), X(k), Y(k)]^T$, coefficient matrices $A_k = e^{A_t \Delta T}$, $B_k = \int_0^{\Delta T} e^{A_t \tau} B_t d\tau$, ΔT denotes the sampling time, control vector $u(k) = [a_x(k), \delta_f(k)]^T$, control increment vector $\Delta u(k) = [\Delta a_x(k), \Delta \delta_f(k)]^T$.

Furthermore, defining a new state vector that combines the original state vector and the control vector.

$$\vartheta(k) = [x(k), u(k-1)]^T \tag{6.4}$$

After that, a new discrete state-space form of (6.3) can be expressed as

$$\begin{cases} \vartheta(k+1) = \tilde{A}_k \vartheta(k) + \tilde{B}_k \Delta u(k) \\ y(k) = \tilde{C}_k \vartheta(k) \end{cases} \tag{6.5}$$

where $\tilde{A}_k = \begin{bmatrix} A_k & B_k \\ 0_{2\times6} & I_2 \end{bmatrix}$, $\tilde{B}_k = \begin{bmatrix} B_k \\ I_2 \end{bmatrix}$ and $\tilde{C}_k = \begin{bmatrix} I_6 & 0_{6\times2} \end{bmatrix}$.

Defining the predictive horizon N_p and the control horizon N_c, $N_p > N_c$. At the time step k, if the state vector $\vartheta(k)$, the control vector$\Delta u(k)$ and coefficient matrices i.e., $\tilde{A}_{p,k}$, $\tilde{B}_{p,k}$ and $\tilde{C}_{p,k}$ are known, the predicted state vectors are derived as follows.

$$\begin{cases} \vartheta(p+1|k) = \tilde{A}_{p,k}\vartheta(p|k) + \tilde{B}_{p,k}\Delta u(p|k) \\ y(p|k) = \tilde{C}_{p,k}\vartheta(p|k) \end{cases} \tag{6.6}$$

where $p = k, k+1, \cdots, k+N_p-1$.

Supposing that $\tilde{A}_{p,k} = \tilde{A}_k$, $\tilde{B}_{p,k} = \tilde{B}_k$ and $\tilde{C}_{p,k} = \tilde{C}_k$, the motion prediction can be obtained.

$$\begin{aligned} \vartheta(k+1|k) &= \tilde{A}_k\vartheta(k|k) + \tilde{B}_k\Delta u(k|k) \\ \vartheta(k+2|k) &= \tilde{A}_k^2\vartheta(k|k) + \tilde{A}_k\tilde{B}_k\Delta u(k|k) + \tilde{B}_k\Delta u(k+1|k) \\ &\vdots \\ \vartheta(k+N_c|k) &= \tilde{A}_k^{N_c}\vartheta(k|k) + \tilde{A}_k^{N_c-1}\tilde{B}_k\Delta u(k|k) + \cdots \\ &\quad + \tilde{B}_k\Delta u(k+N_c-1|k) \\ &\vdots \\ \vartheta(k+N_p|k) &= \tilde{A}_k^{N_p}\vartheta(k|k) + \tilde{A}_k^{N_p-1}\tilde{B}_k\Delta u(k|k) + \cdots \\ &\quad + \tilde{A}_k^{N_p-N_c}\tilde{B}_k\Delta u(k+N_c-1|k) \end{aligned} \tag{6.7}$$

The output vector sequence is defined by

$$\mathbf{Y}(k) = [y^T(k+1|k), y^T(k+2|k), \cdots, y^T(k+N_p|k)]^T \tag{6.8}$$

Based on (6.7) and (6.8), the predicted motion output vector sequence $\mathbf{Y}(k)$ is derived by

$$\mathbf{Y}(k) = \bar{C}\vartheta(k|k) + \bar{D}\Delta\mathbf{u}(k) \tag{6.9}$$

where $\Delta\mathbf{u}(k) = [\Delta u^T(k|k), \Delta u^T(k+1|k), \cdots, \Delta u^T(k+N_c-1|k)]^T$, $\bar{C} = [(\tilde{C}_k\tilde{A}_k)^T, (\tilde{C}_k\tilde{A}_k^2)^T, \cdots, (\tilde{C}_k\tilde{A}_k^{N_p})^T]^T$,

and $\bar{D} = \begin{bmatrix} \tilde{C}_k\tilde{B}_k & 0 & 0 & 0 \\ \vdots & \vdots & \vdots & \vdots \\ \tilde{C}_k\tilde{A}_k^{N_c-1}\tilde{B}_k & \cdots & \tilde{C}_k\tilde{A}_k\tilde{B}_k & \tilde{C}_k\tilde{B}_k \\ \vdots & \vdots & \vdots & \vdots \\ \tilde{C}_k\tilde{A}_k^{N_p-1}\tilde{B}_k & \cdots & \tilde{C}_k\tilde{A}_k^{N_p-N_c+1}\tilde{B}_k & \tilde{C}_k\tilde{A}_k^{N_p-N_c}\tilde{B}_k \end{bmatrix}$.

It can be found that if the control vector sequence $\Delta\mathbf{u}$ for CAVs is known, the motion prediction of CAVs can be realized. $\Delta\mathbf{u}$ can be worked out by solving the decision-making issue of CAVs in the next section.

6.2.3 Decision Making Using the Coalitional Game Approach

This section presents the collaborative decision-making algorithm design of CAVs, resolving the driving conflicts at the multi-lane merging zone. Four kinds of coalition types are introduced firstly. Furthermore, comprehensively considering various driving performances, i.e., driving safety, ride comfort and travel efficiency, the decision-making cost function is designed. Based on the constructed decision-making cost function and multiple constraints, the coalitional game theoretic approach is used for the collaborative decision making of CAVs.

6.2.3.1 Formulation of the Coalitional Game for CAVs

The coalitional game theoretic approach is one of the cooperative game theoretic approaches. All players in the same coalition aim to minimize the cost value of the coalition. The definition of coalitional game is shown as follows.

Definition 6.1 [24]: In the coalitional game, the set of all players is denoted by $N = \{1, 2, \cdots, n\}$, who try to form the coalition to reduce costs. Each subset S of N is called a coalition, i.e., $S \in 2^N$. If S consists of only one player, it can also be seen as a coalition, i.e., a single player coalition. If S is made up of all players, it is called a grand coalition. A coalitional game is defined by a pair $\langle N, U, J \rangle$, where U denotes a set of decision-making behaviors of players, J denotes the characteristic function.

Remark 6.1 : In the game theoretic approach, the characteristic function J is usually represented by a reward function or payoff function. Each player or coalition tries to maximize the reward value. In this study, J is denoted by the cost function. Correspondingly, each player or coalition tries to minimize the cost value.

The characteristic function for the coalition S is denoted by $J^S(U_S)$, $S \in 2^N$. Correspondingly, the characteristic function for a single player coalition is denoted by $J^i(U_i)$, $i \in N$. In the coalitional game, each player can join any coalition, but the decision should follow some rules, shown in the following definition.

Definition 6.2 [16]: Based on the principle of fairness, the individual rationality requires that each player in the coalition should obtain a satisfactory cost allocation which is no more than that without joining the coalition, i.e., $Q_i \leq J^i(U_i)$, $\forall i \in N$, where Q_i denotes the cost allocation of the player i, which is realized by the Shapley allocation rule.

Based on the Definition 6.2, the combination and splitting rules of coalitions are concluded. Any collection of disjoint coalitions S_j, $S_j \in 2^N$, can be combined together to form a single coalition H if and only if

$$\begin{aligned} & J^H(U_H) \leq \sum_{j=1}^{m} J^{S_j}(U_j) \\ & H = S_1 \bigcup S_2 \cdots \bigcup S_m, \ j = \{1, 2, \cdots, m\} \end{aligned} \tag{6.10}$$

Otherwise, the coalition H will split into smaller coalitions.

Remark 6.2: To apply the coalitional game theoretic approach to the collaborative decision making of CAVs at the multi-lane merging zone, we should give the definition of player, decision-making behavior and the characteristic function. Not all CAVs are considered as players. In Figure 6.1, only the CAVs in the merging conflict zone are regarded as players, i.e., V1, V2 and V3. The decision-making behavior includes the merging behavior, acceleration or acceleration behavior and the lane-change behavior. In this study, the merging behavior and the lane-change behavior are combined into one behavior. Besides, the characteristic function is represented by the cost function in this study. The detailed construction process is presented in the following subsection.

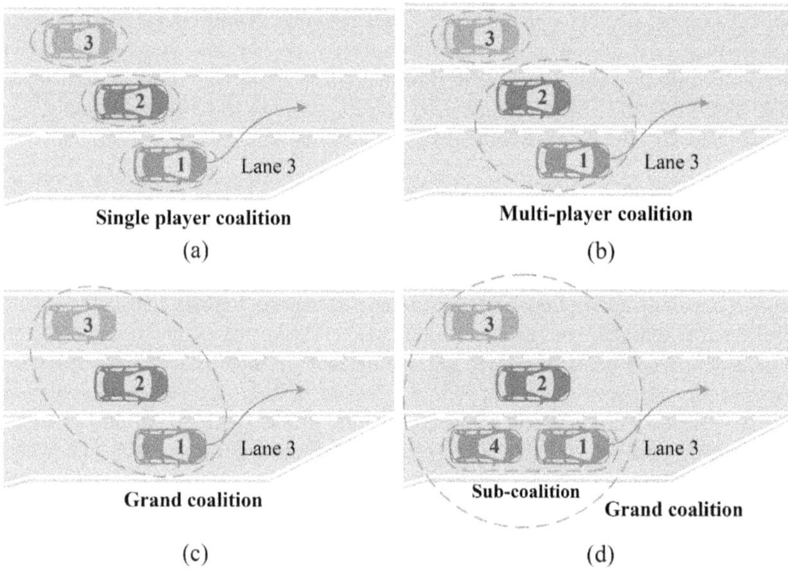

FIGURE 6.3: Four types of coalition for CAVs at the multi-lane merging zone: (a) the single player coalition; (b) the multi-player coalition; (c) the grand coalition; (d) the grand coalition with a sub-coalition.

In the coalitional game theoretic approach, CAVs can form different kinds of coalitions. Figure 6.3 shows four kinds of typical coalitions, i.e., single player coalition, multi-player coalition, grand coalition, and sub-coalition. Figure 6.3(a) shows the single player coalition. Each CAV forms an independent coalition. Namely, each CAV is unwilling to cooperate with other CAVs, which is an extreme condition and can be regarded as a noncooperative game. Figure 6.3(b) shows the multi-player coalition, in which V1 and V2 form a two-player coalition to realize cooperative decision making. However, V3 forms a single player coalition. The relationship between V3 and the two-player coalition is noncooperative. Besides the single player coalition, the grand coalition is another extreme coalition type, shown in Figure 6.3(c). We can see that all CAVs are the multi-lane merging zone form a grand coalition. Namely, the relationship between CAVs in the grand coalition is cooperative. There is no competitive relationship in the grand coalition. Figure 6.3(d) shows that the grand coalition includes a sub-coalition. Due to the same driving objective and platoon control, V1 and V4 form a sub-coalition in the grand coalition. Namely, V1 and V4 can be regarded as one player. To sum up, all the four coalition types can exist in the real coalitional game. The coalition combination should follow the rules described in Definition 6.2. It is worth mentioning that the above coalition types are the father, which can be generalized to more coalition types in real applications.

6.2.3.2 Cost Function for the Decision Making of an Individual CAV

In Figure 6.1, V1 is ready to merge into the main road from the on-ramp lane. V1's decision-making behavior includes the longitudinal decision making, e.g., acceleration and deceleration, and the lateral decision making, e.g., lane-change and lane-keeping. Correspondingly, V2 has the similar decision-making behavior to address the merging conflict of V1. Since V3 is on the left lane, we only consider the longitudinal decision-making behavior of V3, i.e., acceleration and deceleration. We can find the longitudinal decision making is mainly affected by the lead vehicle (LV), and the lateral decision making is mainly related to the neighbour vehicle (NV).

In the decision-making cost function construction of CAVs, three critical driving performances are considered, i.e., driving safety, ride comfort and travel efficiency. If Vi is regarded as the host vehicle (HV), the decision-making cost function can be written as

$$J^{Vi} = \omega_s^{Vi} J_s^{Vi} + \omega_c^{Vi} J_c^{Vi} + \omega_e^{Vi} J_e^{Vi} \tag{6.11}$$

where J_s^{Vi}, J_c^{Vi} and J_e^{Vi} are the cost functions regarding driving safety, ride comfort and travel efficiency, respectively. ω_s^{Vi}, ω_c^{Vi} and ω_e^{Vi} denote the weighting coefficients, reflecting the driving preference. Referring to [147, 149], the weighting coefficients regarding three kinds of driving styles are presented in Table 6.1.

The cost function regarding driving safety J_s^{Vi} includes the longitudinal safety, lateral safety and lane keeping safety.

$$\begin{aligned} J_s^{Vi} = & ((\beta^{Vi})^2 - 1)^2 J_{s-log}^{Vi} + (\beta^{Vi})^2 J_{s-lat}^{Vi} \\ & + ((\beta^{Vi})^2 - 1)^2 J_{s-lk}^{Vi} + J_{s-lc}^{Vi} \end{aligned} \tag{6.12}$$

where J_{s-log}^{Vi}, J_{s-lat}^{Vi}, J_{s-lk}^{Vi} and J_{s-lc}^{Vi} are the cost functions regarding the longitudinal, lateral, lane-keeping and lane-change safety, respectively. β^{Vi} denotes the lane-change behavior, $\beta^{Vi} \in \{-1, 0, 1\} :=$ {*changing lanes to the left, lane keeping, changing lanes to the right*}.

TABLE 6.1: Weighting Coefficients of Different Driving Characteristics

Driving Characteristic	Weighting Coefficients		
	ω_s^{Vi}	ω_c^{Vi}	ω_e^{Vi}
Aggressive	0.1	0.1	0.8
Moderate	0.5	0.3	0.2
Conservative	0.7	0.2	0.1

The cost function regarding the longitudinal safety J_{s-log}^{Vi} is related to the relative distance and relative velocity between Vi and its LV, derived by

$$J_{s-log}^{Vi} = \varpi_{v-log}^{Vi} \eta_\sigma^{Vi} (\Delta v_{x,\sigma}^{Vi})^2 + \varpi_{s-log}^{Vi} / [(\Delta s_\sigma^{Vi})^2 + \varepsilon] \tag{6.13}$$

$$\Delta v_{x,\sigma}^{Vi} = v_{x,\sigma}^{LV} - v_{x,\sigma}^{Vi} \tag{6.14}$$

$$\Delta s_\sigma^{Vi} = [(X_\sigma^{LV} - X_\sigma^{Vi})^2 + (Y_\sigma^{LV} - Y_\sigma^{Vi})^2]^{1/2} - L_V \tag{6.15}$$

$$\eta_\sigma^{Vi} = 0.5 - 0.5\text{sgn}(\Delta v_{x,\sigma}^{Vi}) \tag{6.16}$$

where $v_{x,\sigma}^{LV}$ and $v_{x,\sigma}^{Vi}$ are the longitudinal velocities of LV and Vi, respectively. $(X_\sigma^{LV}, Y_\sigma^{LV})$ and $(X_\sigma^{Vi}, Y_\sigma^{Vi})$ denote the positions of LV and Vi, respectively. ϖ_{v-log}^{Vi} and ϖ_{s-log}^{Vi} denote the weighting coefficients. ε is a design parameter of a small value to avoid zero denominator in the calculation. L_V denotes a safety coefficient considering the vehicle length. σ is the lane number labelled from left to right, $\sigma \in \{1, 2, 3\}$:=$\{$*lane 1, lane 2, lane 3*$\}$. η_σ^{Vi} is a switch function. If $\Delta v_{x,\sigma}^{Vi} > 0$, i.e., $v_{x,\sigma}^{LV} > v_{x,\sigma}^{Vi}$, $\eta_\sigma^{Vi} = 0$.

The cost function of the lateral safety J_{s-lat}^{Vi} is related to the relative distance and relative velocity between Vi and its NV, expressed by

$$J_{s-lat}^{Vi} = \varpi_{v-lat}^{Vi} \eta_{\sigma+\beta^{Vi}}^{Vi} (\Delta v_{x,\sigma+\beta^{Vi}}^{Vi})^2 + \varpi_{s-lat}^{Vi} / [(\Delta s_{\sigma+\beta^{Vi}}^{Vi})^2 + \varepsilon] \tag{6.17}$$

$$\Delta v_{x,\sigma+\beta^{Vi}}^{Vi} = v_{x,\sigma}^{Vi} - v_{x,\sigma+\beta^{Vi}}^{NV} \tag{6.18}$$

$$\Delta s_{\sigma+\beta^{Vi}}^{Vi} = [(X_\sigma^{Vi} - X_{\sigma+\beta^{Vi}}^{NV})^2 + (Y_\sigma^{Vi} - Y_{\sigma+\beta^{Vi}}^{NV})^2]^{1/2} - L_V \tag{6.19}$$

$$\eta_{\sigma+\beta^{Vi}}^{Vi} = 0.5 - 0.5\text{sgn}(\Delta v_{x,\sigma+\beta^{Vi}}^{Vi}) \tag{6.20}$$

where $v_{x,\sigma+\beta^{Vi}}^{NV}$ denotes the longitudinal velocity of NV. $(X_{\sigma+\beta^{Vi}}^{NV}, Y_{\sigma+\beta^{Vi}}^{NV})$ denotes the position of NV. ϖ_{v-lat}^{Vi} and ϖ_{s-lat}^{Vi} denote the weighting coefficients. $\eta_{\sigma+\beta^{Vi}}^{Vi}$ is a switch function. If $\Delta v_{x,\sigma+\beta^{Vi}}^{Vi} > 0$, i.e., $v_{x,\sigma}^{Vi} > v_{x,\sigma+\beta^{Vi}}^{NV}$, $\eta_{\sigma+\beta^{Vi}}^{Vi} = 0$.

The cost function regarding the lane-keeping safety J_{s-lk}^{Vi} is associated with the lateral distance error and the yaw angle error between the predicted position of Vi and the center line of the lane σ .

$$J_{s-lk}^{Vi} = \varpi_{y-lk}^{Vi} (\Delta y_\sigma^{Vi})^2 + \varpi_{\varphi-lk}^{Vi} (\Delta \varphi_\sigma^{Vi})^2 \tag{6.21}$$

where Δy_σ^{Vi} and $\Delta \varphi_\sigma^{Vi}$ denote the lateral distance error and yaw angle error, ϖ_{y-lk}^{Vi} and $\varpi_{\varphi-lk}^{Vi}$ denote the weighting coefficients.

The cost function regarding the lane-change safety J_{s-lc}^{Vi} is used to enhance the driving safety during the lane change process, and it can be calculated based on a potential field model [78].

$$J_{s-lc}^{Vi} = \Gamma^{LV} + \Gamma^{NV} \tag{6.22}$$

$$\Gamma^\jmath = \hbar^\jmath e^\Psi, \quad (\jmath = LV, NV) \tag{6.23}$$

$$\Psi = -\{\frac{\hat{X}^2}{2\sigma_X^2} + \frac{\hat{Y}^2}{2\sigma_Y^2}\}^\varrho + \varsigma v_x^\jmath \Upsilon \tag{6.24}$$

$$\Upsilon = k^{\jmath} \frac{\hat{X}^2}{2\sigma_X^2} / \sqrt{\frac{\hat{X}^2}{2\sigma_X^2} + \frac{\hat{Y}^2}{2\sigma_Y^2}} \tag{6.25}$$

$$k^{\jmath} = \begin{cases} -1, \hat{X} < 0 \\ 1, \quad \hat{X} \geq 0 \end{cases} \tag{6.26}$$

$$\begin{bmatrix} \hat{X} \\ \hat{Y} \end{bmatrix} = \begin{bmatrix} \cos\varphi^{\jmath} & \sin\varphi^{\jmath} \\ -\sin\varphi^{\jmath} & \cos\varphi^{\jmath} \end{bmatrix} \begin{bmatrix} X - X^{\jmath} \\ Y - Y^{\jmath} \end{bmatrix} \tag{6.27}$$

where $\Gamma^{\jmath}$ is the potential field value induced by LV or NV at the position (X, Y). $(X^{\jmath}, Y^{\jmath})$ are the positions of LV and NV. $\varphi^{\jmath}$ and $v_x^{\jmath}$ are the yaw angle and velocity of LV and NV. σ_X and σ_Y denote the convergence coefficients. $\hbar$, ϱ and ς denote the shape coefficients.

The cost function of ride comfort J_c^{Vi} is related to the jerk of Vi.

$$J_c^{Vi} = \varpi_{jx}^{Vi}(j_{x,\sigma}^{Vi})^2 + \varpi_{jy}^{Vi}(j_{y,\sigma}^{Vi})^2 \tag{6.28}$$

where $j_{x,\sigma}^{Vi}$ and $j_{y,\sigma}^{Vi}$ denote the longitudinal and lateral jerks of Vi. ϖ_{jx}^{Vi} and ϖ_{jy}^{Vi} denote the weighting coefficients.

Besides, the cost function of travel efficiency J_e^{Vi} is related to the longitudinal velocity of Vi, which is derived by

$$J_e^{Vi} = \varpi_e^{Vi}(v_{x,\sigma}^{Vi} - \hat{v}_{x,\sigma}^{Vi})^2 \tag{6.29}$$

$$\hat{v}_{x,\sigma}^{Vi} = \min(v_{x,\sigma}^{\max}, v_{x,\sigma}^{LV}) \tag{6.30}$$

where $v_{x,\sigma}^{\max}$ denotes the velocity limit on the lane σ, and ϖ_e^{Vi} denotes the weighting coefficient.

6.2.3.3 Constraints of the Decision Making

To guarantee the decision-making safety, multiple constraints are considered in the decision-making algorithm. The safety constraints for Vi are given by

$$|\Delta s_\sigma^{Vi}| \leq \Delta s^{\max}, |\Delta y_\sigma^{Vi}| \leq \Delta y^{\max}, |\Delta \varphi_\sigma^{Vi}| \leq \Delta \varphi^{\max} \tag{6.31}$$

The constraints for ride comfort are expressed as

$$|j_{x,\sigma}^{Vi}| \leq j_x^{\max}, |j_{y,\sigma}^{Vi}| \leq j_y^{\max} \tag{6.32}$$

The constraints for accelerations are expressed as

$$|a_{x,\sigma}^{Vi}| \leq a_x^{\max}, |a_{y,\sigma}^{Vi}| \leq a_y^{\max} \tag{6.33}$$

Besides, the constraint for travel efficiency is given by

$$|v_{x,\sigma}^{Vi}| \leq v_{x,\sigma}^{\max} \tag{6.34}$$

Moreover, the constraint of the curvature trajectory is derived by

$$\frac{|\dot{X}_\sigma^{Vi}\ddot{Y}_\sigma^{Vi} - \ddot{X}_\sigma^{Vi}\dot{Y}_\sigma^{Vi}|}{[(\dot{X}_\sigma^{Vi})^2 + (\dot{Y}_\sigma^{Vi})^2]^{3/2}} \leq \frac{1}{R_{\min}} \tag{6.35}$$

$$\dot{X}_\sigma^{Vi} = [X_\sigma^{Vi}(k+1) - X_\sigma^{Vi}(k)]/\Delta T \tag{6.36}$$

$$\dot{Y}_\sigma^{Vi} = [Y_\sigma^{Vi}(k+1) - Y_\sigma^{Vi}(k)]/\Delta T \tag{6.37}$$

$$\ddot{X}_\sigma^{Vi} = [X_\sigma^{Vi}(k+2) - 2X_\sigma^{Vi}(k+1) + X_\sigma^{Vi}(k)]/\Delta T^2 \tag{6.38}$$

$$\ddot{Y}_\sigma^{Vi} = [Y_\sigma^{Vi}(k+2) - 2Y_\sigma^{Vi}(k+1) + Y_\sigma^{Vi}(k)]/\Delta T^2 \tag{6.39}$$

where $R_{\min}$ is the minimum turning radius.

The constraint of the steering angle is given by

$$|\delta_f^{Vi}| \le \delta_f^{\max}, \quad |\Delta\delta_f^{Vi}| \le \Delta\delta_f^{\max} \tag{6.40}$$

The constraint of Δa_x^{Vi} is defined by

$$|\Delta a_x^{Vi}| \le \Delta a_x^{\max} \tag{6.41}$$

Finally, all the constraints for Vi are written in a compact form, shown as follows.

$$\begin{aligned} &\Phi^{Vi}(\Delta s_\sigma^{Vi}, \Delta y_\sigma^{Vi}, \Delta\varphi_\sigma^{Vi}, a_{x,\sigma}^{Vi}, a_{y,\sigma}^{Vi}, j_{x,\sigma}^{Vi}, \\ &j_{y,\sigma}^{Vi}, v_{x,\sigma}^{Vi}, X_\sigma^{Vi}, Y_\sigma^{Vi}, \delta_f^{Vi}, \Delta\delta_f^{Vi}, \Delta a_x^{Vi}) \end{aligned} \tag{6.42}$$

6.2.3.4 Decision Making with the Coalitional Game Approach

Considering that the motion prediction module can advance the accuracy of the decision-making algorithm, the MPC approach is applied to the prediction of the decision-making cost functions of CAVs.

At the time step k, the cost function sequence of Vi is expressed as follows.

$$J^{Vi}(k+1|k), J^{Vi}(k+2|k), \cdots, J^{Vi}(k+N_p|k) \tag{6.43}$$

Besides, the decision-making sequence of Vi is derived by

$$\hat{u}^{Vi}(k|k), \hat{u}^{Vi}(k+1|k), \cdots, \hat{u}^{Vi}(k+N_c-1|k) \tag{6.44}$$

where $\hat{u}^{Vi}(q|k) = [\Delta a_x^{Vi}(q|k), \Delta\delta_f^{Vi}(q|k), \beta^{Vi}(q|k)]^T$, $q = k, k+1, \cdots, k+N_c-1$.

Furthermore, the characteristic function of Vi is given by

$$\Lambda^{Vi} = \sum_{p=k+1}^{k+N_p} ||J^{Vi}(p|k)||_Q^2 + \sum_{q=k}^{k+N_c-1} ||\hat{u}^{Vi}(q|k)||_R^2 \tag{6.45}$$

where Q and R denote the weighting matrices.

As mentioned above, four kinds of coalition types are proposed for CAVs at the multi-lane merging zone. Correspondingly, the following four decision-making strategies are derived.

(1) The single player coalition:

As Figure 6.3(a) shows, each CAV forms an independent single player coalition, i.e., $S_1 = \{V1\}$, $S_2 = \{V2\}$, $S_3 = \{V3\}$. The decision-making sequences for the three CAVs are expressed as

$$(\Delta a_x^{V1*}, \Delta \delta_f^{V1*}, \beta^{V1*}) = \arg\min \Lambda^{V1} \tag{6.46}$$

$$(\Delta a_x^{V2*}, \Delta \delta_f^{V2*}, \beta^{V2*}) = \arg\min \Lambda^{V2} \tag{6.47}$$

$$\Delta a_x^{V3*} = \arg\min \Lambda^{V3} \tag{6.48}$$

s.t. $\Phi^{V1} \leq 0$, $\Phi^{V2} \leq 0$, $\Phi^{V3} \leq 0$, $\beta^{V1}(\beta^{V1}+1) = 0$, $\beta^{V2}(\beta^{V2}+1) = 0$.
where $\Delta a_x^{Vi*}, \Delta \delta_f^{Vi*}$ and β^{Vi*} are the optimal decision-making sequence of Vi.

(2) The multi-player coalition:

As Figure 6.3(b) shows, V1 and V2 form a two-player coalition, and V3 is a single player coalition, i.e., $S_1 = \{V1, V2\}$, $S_2 = \{V3\}$. The decision-making sequences of the two coalitions are derived by

$$\begin{aligned}&(\Delta a_x^{V1*}, \Delta \delta_f^{V1*}, \beta^{V1*}, \Delta a_x^{V2*}, \Delta \delta_f^{V2*}, \beta^{V2*}) \\ &= \arg\min[\Lambda^{V1} + \Lambda^{V2}]\end{aligned} \tag{6.49}$$

$$\Delta a_x^{V3*} = \arg\min \Lambda^{V3} \tag{6.50}$$

s.t. $\Phi^{V1} \leq 0$, $\Phi^{V2} \leq 0$, $\Phi^{V3} \leq 0$, $\beta^{V1}(\beta^{V1}+1) = 0$, $\beta^{V2}(\beta^{V2}+1) = 0$.

(3) The grand coalition:

As Figure 6.3(c) shows, V1, V2 and V3 form a grand coalition, i.e., $S_1 = \{V1, V2, V3\}$. The decision-making sequence is expressed as follows.

$$\begin{aligned}&(\Delta a_x^{V1*}, \Delta \delta_f^{V1*}, \beta^{V1*}, \Delta a_x^{V2*}, \Delta \delta_f^{V2*}, \beta^{V2*}, \\ &\Delta a_x^{V3*}) = \arg\min[\Lambda^{V1} + \Lambda^{V2} + \Lambda^{V3}]\end{aligned} \tag{6.51}$$

s.t. $\Phi^{V1} \leq 0$, $\Phi^{V2} \leq 0$, $\Phi^{V3} \leq 0$, $\beta^{V1}(\beta^{V1}+1) = 0$, $\beta^{V2}(\beta^{V2}+1) = 0$.

(4) The grand coalition including a sub-coalition:

As Figure 6.3(d) shows, all CAVs form a grand coalition, in which there exists a sub-coalition, i.e., $S_1 = \{\{V1, V4\}, V2, V3\}$. To simplify the complexity of the decision-making algorithm, V1 and V4 are regarded as one player in the game. As a result, V1 and V4 have the same decision-making behavior, but there exists a time delay of V4's behavior. The decision-making sequence is expressed by

$$\begin{aligned}&(\Delta a_x^{V1*}, \Delta \delta_f^{V1*}, \beta^{V1*}, \Delta a_x^{V2*}, \Delta \delta_f^{V2*}, \beta^{V2*}, \\ &\Delta a_x^{V3*}) = \arg\min[\Lambda^{V1} + \Lambda^{V2} + \Lambda^{V3}]\end{aligned} \tag{6.52}$$

$$\begin{aligned}\Delta a_x^{V4*}(k) &= \Delta a_x^{V1*}(k-\tau) \\ \Delta \delta_f^{V4*}(k) &= \Delta \delta_f^{V1*}(k-\tau) \\ \beta^{V4*}(k) &= \beta^{V1*}(k-\tau)\end{aligned} \tag{6.53}$$

s.t. $\Phi^{V1} \leq 0$, $\Phi^{V2} \leq 0$, $\Phi^{V3} \leq 0$, $\beta^{V1}(\beta^{V1}+1) = 0$, $\beta^{V2}(\beta^{V2}+1) = 0$.

Algorithm 1 Decision-making algorithm for Vi at the multi-lane merging zone.

Input the motion information of Vi (on the Lane 3);
Input the motion information of NVs and LV for Vi. Vj and Vk are the NVs for Vi on the Lane 2 and Lane 1;
for $i = 1 : n$ **do**
    Sub-coalition formation with Algorithm 2;
    Coalition $S = \{Si, Sj, Sk\}$;
    for $\xi = 1 : length(S)$ **do**
        if $Q_{S(\xi)} > J^{S(\xi)}(U_{S(\xi)})$ **then**
            $S(\xi)$ breaks away from S;
        else
            $S(\xi)$ remains in S;
        end if
    end for
    if only one sub-coalition, $S(\xi)$, breaks away **then**
        if $S(\xi)$=Si **then**
            S splits into $\{Si\}$, $\{Sj,\ Sk\}$;
        else
            if $S(\xi)$=Sj **then**
                S splits into $\{Sj\}$, $\{Si,\ Sk\}$;
            else
                S splits into $\{Sk\}$, $\{Si,\ Sj\}$;
            end if
        end if
    else
        if Two or more sub-coalitions break away **then**
            S splits into $\{Si\}$, $\{Sj\}$, $\{Sk\}$;
        else
            No splitting, $S = \{Si, Sj, Sk\}$;
        end if
    end if
    Optimization with Eq.35-Eq.38;
    Output the decision-making results.
end for

where τ is the decision-making delay of V4 following V1, which is given by $\tau = \Delta s_{\sigma}^{V4}/v_{x,\sigma}^{V4}/\Delta T$.

Algorithm 1 and Algorithm 2 show the formation principle and process of the coalitions for CAVs at the merging conflict zone. From Algorithm 1, we can find that all CAVs that enter the merging zone will be assigned with a number firstly. Vi on the Lane 3 wants to merge into the Lane 2, Vk and Vj on the Lane 1 and Lane 2 are the NVs for Vi. Then, the sub-coalition formation is done based on Algorithm 2. As a result, three sub-coalitions are created in Lane 1, Lane 2 and Lane3, respectively, denoted by Si, Sj, Sk. o find the optimal coalition formation, the grand coalition S is

Algorithm 2 Algorithm for sub-coalition formation.

Coalition $Si = \{Vi\}$;
for $\zeta = i + 1 : n$ **do**
    if $\Delta s^{V\zeta} < \Delta s_0$ and $\omega^{V\zeta} = \omega^{V\zeta-1}$ **then**
        $Si \leftarrow add \quad V\zeta$
    else
        Break;
    end if
end for
Coalition $Sj = \{Vj\}$;
for $\eta = j + 1 : p$ **do**
    if $\Delta s^{V\eta} < \Delta s_0$ and $\omega^{V\eta} = \omega^{V\eta-1}$ **then**
        $Sj \leftarrow add \quad V\eta$
    else
        Break;
    end if
end for
Coalition $Sk = \{Vk\}$;
for $\gamma = k + 1 : q$ **do**
    if $\Delta s^{V\gamma} < \Delta s_0$ and $\omega^{V\gamma} = \omega^{V\gamma-1}$ **then**
        $Sk \leftarrow add \quad V\gamma$
    else
        Break;
    end if
end for
Output Si, Sj, Sk.

created at first, i.e., $S = \{Si, Sj, Sk\}$. Then, the sub-coalitions choose to either break away from S or remain in S according to the principle in Definition 2. Therefore, the grand coalition S will split into different coalition types, which are covered by the proposed four types of coalitions. Finally, the decision-making results of all CAVs are worked out by solving the optimization issue formulated by (6.46) to (6.53).

It can be found that the game-theoretic decision-making issue is finally transformed into a closed-loop iterative optimization process with multiple constraints [285], which is solved with the efficient evolutionary algorithm based on convex optimization theory [200].

6.2.4 Testing, Validation and Discussion

To evaluate the performance of the coalitional game-theoretic decision-making algorithm, two test cases are designed and carried out with the MATLAB/Simulink simulation platform. The motion planning and control algorithms are adapted from Chapter 4. Referring to [79, 76], the simulation parameters for decision making are shown in Table 6.2.

TABLE 6.2: Parameters of the Decision-Making Algorithm

Parameter	Value	Parameter	Value
$\Delta s^{\max}$/ (m)	0.8	$j_y^{\max}$/ (m/s^3)	2
$\Delta y^{\max}$/ (m)	0.2	$v_{x,\sigma}^{\max}$/ (m/s)	30
$\Delta\varphi^{\max}$/ (deg)	2	$R_{\min}$/ (m)	8
$a_x^{\max}$/ (m/s^2)	4	$\delta_f^{\max}$/ (deg)	30
$a_y^{\max}$/ (m/s^2)	4	$\Delta\delta_f^{\max}$/ (deg)	0.3
$j_x^{\max}$/ (m/s^3)	2	$\Delta a_x^{\max}$/ (m/s^2)	0.1
N_p	5	N_c	2

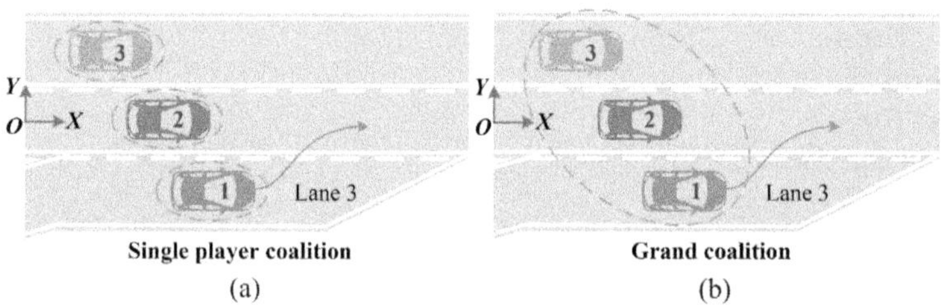

FIGURE 6.4: Comparison of the two coalitional types considering different driving characteristics of CAVs: (a) the single player coalition; (b) the grand coalition.

6.2.4.1 Case Study 1

The test scenario of Case 1 is illustrated in Figure 6.4 which includes three CAVs at the merging conflict zone. This case aims to conduct the comparative study between two extreme coalitions, i.e., the single player coalition and the grand coalition. Essentially, it is a comparative test between the noncooperative game and the cooperative game. Moreover, the effects of different driving styles on the two coalitions are evaluated. The initial positions of V1, V2, V3, and the LVs of V1, V2 and V3 are set as (12, -4), (10, 0), (8, 4), (62, -4), (70, 0) and (68, 4), respectively. Besides, the initial velocities of V1, V2, V3, and the LVs of V1, V2 and V3 are set as 18 m/s, 19 m/s, 20 m/s, 26 m/s, 26 m/s, 26 m/s, respectively.

Additionally, two scenarios are designed in this case considering the personalized driving of CAVs. In Scenario A, the driving styles of the three CAVs are set as normal. In Scenario B, the driving styles of V2 and V3 are set as normal, and the driving style of V1 is set as aggressive.

Figure 6.5 shows the test results in the two scenarios. In Scenario A and Scenario B, V1, V2 and V3 made the similar decisions. V1 changed lanes and merged into the main lane successfully. To resolve the merging conflict of V1, V2 changed lanes to the left lane and V3 slowed down to give ways to V2. Besides, both Scenario A and Scenario B show that the merging time of the single player coalition (Coalition

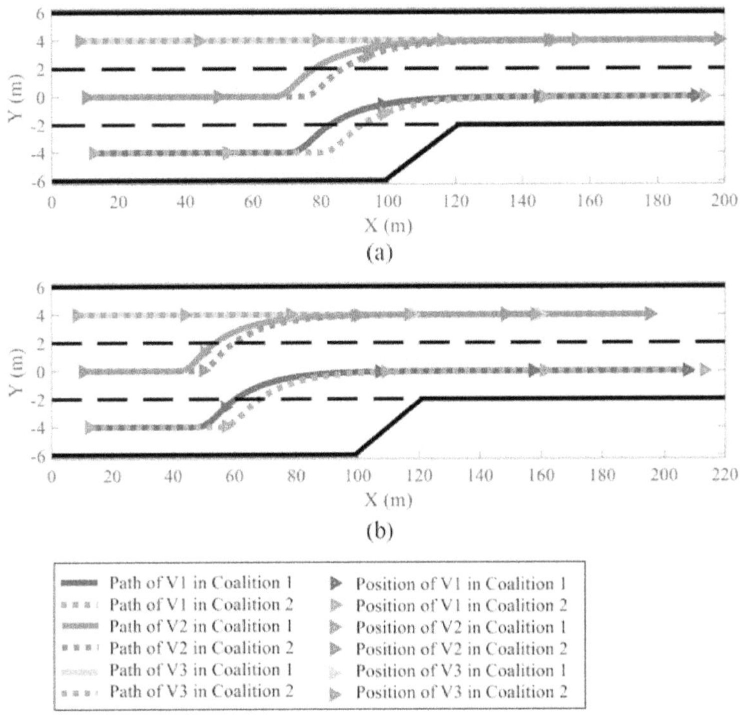

FIGURE 6.5: Decision-making results of the two scenarios considering different driving characteristics of CAVs in Case 1: (a) Scenario A; (b) Scenario B.

TABLE 6.3: Cost Values of CAVs in Case 1 with Different Scenario Settings

Cost Values (RMS)	Scenario A		Scenario B	
	Single Player Coalition	Grand Coalition	Single Player Coalition	Grand Coalition
V1	62598	62196	71527	72903
V2	55560	55086	68933	67071
V3	48521	47983	65914	64297
Sum	166679	165265	206374	204271

1) is earlier than that of the grand coalition (Coalition 2). The cost values of CAVs in Case 1 are shown in Table 6.3. We can see the detailed decision-making results of the two kinds of coalitions. In Scenario A, compared with the single player coalition, the decision-making cost values of all CAVs are reduced in the grand coalition, indicating that, in Scenario A, the grand coalition is beneficial to both the individual

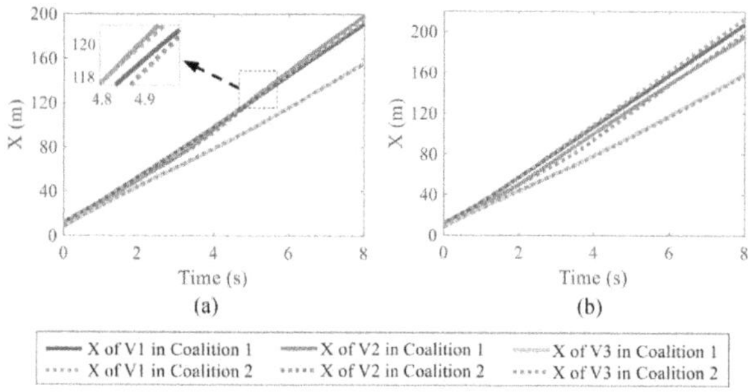

FIGURE 6.6: Longitudinal paths of CAVs in the two scenarios considering different driving characteristics in Case 1: (a) Scenario A; (b) Scenario B.

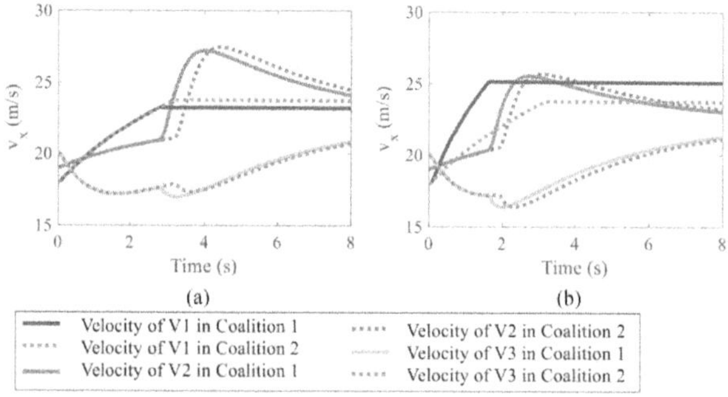

FIGURE 6.7: Velocities of CAVs in the two scenarios considering different driving characteristics in Case 1: (a) Scenario A; (b) Scenario B.

interest and the coalition interest. However, Scenario B shows the different test results regarding the decision-making cost value. The decision-making cost values of V2, V3 and the sum are decreased in the grand coalition, but the decision-making cost value of V1 is increased. The main reason is the different driving style of V1 in Scenario A and Scenario B. In Scenario B, the driving style of V1 is aggressive rather than normal. As a result, V1 pursues higher travel efficiency, which is different from the preference of V2 and V3. In the grand coalition, all CAVs aim to minimize the cost value of the entire grand coalition rather than the individual. However, in the single player coalition, each CAV only cares about the individual interest. Therefore, compared with the single player coalition, the decision-making cost value of V1 is increased in the grand coalition, but the coalition cost is reduced. It can be concluded that the two kinds of coalitions have different characteristics and can be used for different objectives.

Besides, Figures 6.6 and 6.7 show the test results of the longitudinal paths and velocities of CAVs. It can be found that different driving styles lead to different decision-making results. In Scenario B, the aggressive driving style of V1 leads to a sudden acceleration, which is different from the driving behavior of V1 in Scenario A. It indicates that the aggressive driving style prefers higher travel efficiency than

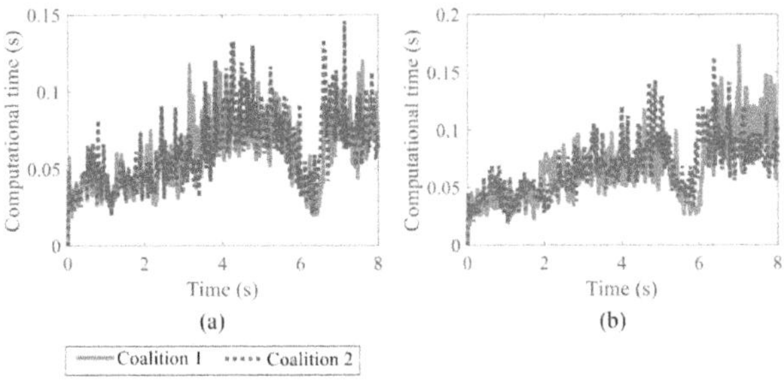

FIGURE 6.8: Computational time in Case 1: (a) Scenario A; (b) Scenario B.

the normal driving style. Moreover, we can see the velocity difference between the two coalitions. In Scenario B, compared with the single player coalition, the velocity of V1 is decreased in the grand coalition. However, the velocities of V2 and V3 are increased in the grand coalition. It indicates that the aggressive driving style is weakened in the grand coalition, but the travel efficiency of the entire traffic system is improved. For personalized driving, the single player coalition is the optimal choice. For the improvement of the traffic system efficiency, the grand coalition is better. This case aims to conduct the comparative study of the two coalitions. Each CAV can choose different coalitions independently, which should follow the coalitional rules in the Definition 6.2.

To analyze the algorithm complexity, the computational time of the proposed algorithm in Case 1 is illustrated in Figure 6.8. The mean value of each step is about 0.06s. The computational efficiency can be further improved with the efficient solver and algorithm improvement in the future.

6.2.4.2 Case Study 2

In this case, the effects of different driving styles on the coalition formation are studied. Figure 6.9 presents the test case, which includes five CAVs at the multi-lane merging zone. Different driving styles are given to the five CAVs. In Scenario A, the driving styles of the five CAVs are set as normal. In Scenario B, the driving styles of V1, V2 and V4 are set as normal, and the driving styles of V3 and V5 are set as aggressive. In Scenario C, the driving style of V2 is set as conservative, and the driving styles of V1, V3, V4 and V5 are set as normal.

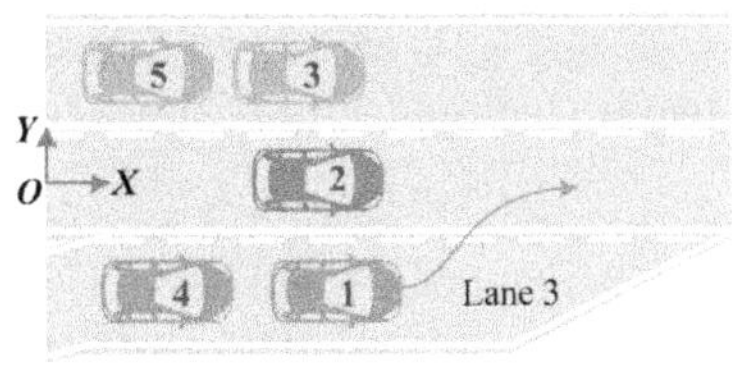

FIGURE 6.9: The decision making of five CAVs at the multi-lane merging zone.

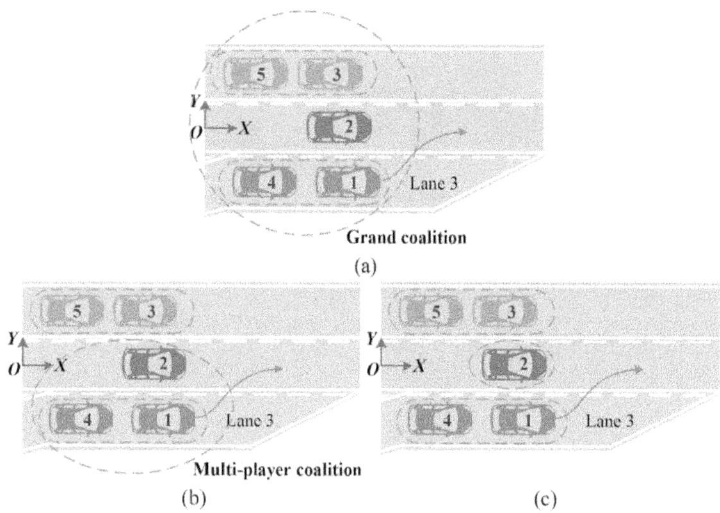

FIGURE 6.10: The coalition formation results of CAVs in Case 2: (a) Scenario A; (b) Scenario B; (c) Scenario C.

In Case 2, the initial positions of V1, V2, V3, V4, V5 and the LVs of V1, V2 and V3 are set as (18, -4), (10, 0), (8, 4), (12, -4), (2, 4), (62, -4), (70, 0) and (68, 4), respectively. The initial velocities of V1, V2, V3, V4, V5 and the LVs of V1, V2 and V3 are set as 20 m/s, 22 m/s, 16 m/s, 20 m/s, 16 m/s, 28 m/s, 28 m/s, 22 m/s, respectively. Finally, the test results are illustrated in Figures 6.10 to 6.14.

From Figure 6.10, it can be found that different driving styles of CAVs lead to different coalition formations. In Scenario A, all CAVs join into a grand coalition, i.e., $S_1 = \{V1, V2, V3, V4, V5\}$, in which V1 and V4 form a sub-coalition, and V3 and V5 form another sub-coalition as well. $S_1 \Rightarrow \{\{V1, V4\}, V2, \{V3, V5\}\}$. From Figure 6.11(a), we can see that, in the grand coalition, V1 and V4 changed lanes and merged into the main lane successfully. V2 changed lanes to the left lane and gave ways for the merging of V1 and V4. As a result, V3 and V5 slowed down to cooperate with V2, in favor of the merging of V1 and V4. In Scenario B, due to the aggressive driving style of V3 and V5, their sub-coalition breaks away from the grand coalition. Finally, V1, V2 and V4 form another multi-player coalition, i.e., $S_1 = \{\{V1, V4\}, V2\}$, $S_2 = \{V3, V5\}$. From Figure 6.11(b), we can see that V3 and V5 are unwilling to give ways for V2. As a result, V2 had to slow down to provide larger safe distance for the merging behavior of V1 and V4. In Scenario C, due to the conservative driving style of V2, V2 forms a single player coalitions. V1 and V4 form a multi-player coalition. V3 and V5 form a multi-player coalition, i.e., $S_1 = \{V1, V4\}$, $S_2 = \{V2\}$, $S_3 = \{V3, V5\}$. Since the driving style of V2 is conservative, to guarantee the driving safety, V2 can only make the deceleration decision to give ways for the merging of V1 and V4.

Figures 6.12 and 6.13 show the longitudinal paths and velocities of CAVs in Case 2. We can find that different driving styles result in different decision-making results of velocity. The aggressive driving style has larger travel velocity than the normal and conservative styles, indicating that the aggressive driving style gives

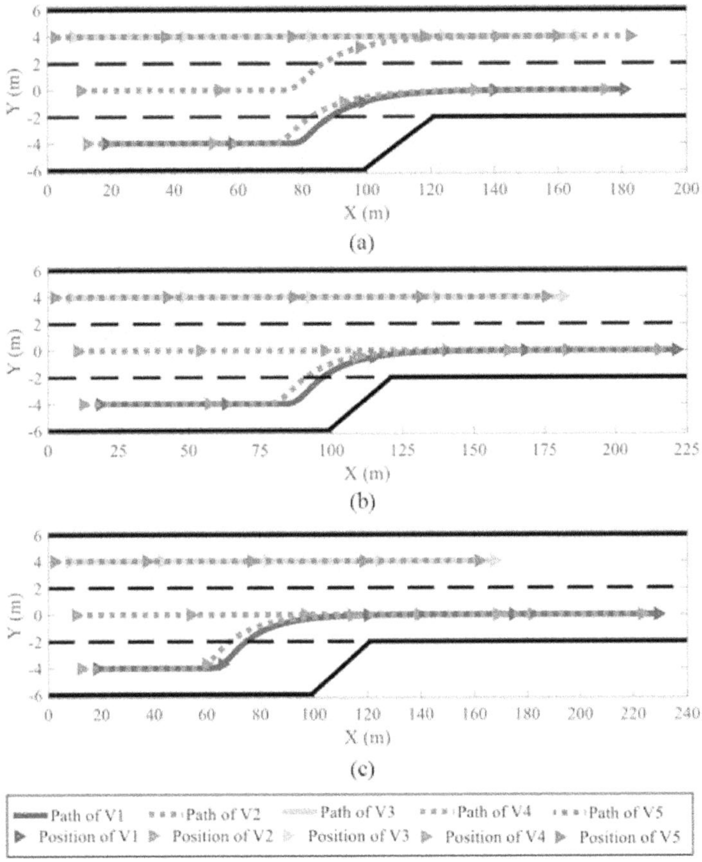

FIGURE 6.11: Decision-making results of the three scenarios in Case 2: (a) Scenario A; (b) Scenario B; (c) Scenario C.

more priority on travel efficiency. However, the conservative driving style cares more about driving safety rather than travel efficiency.

Additionally, Figure 6.14 shows the computational time of the proposed algorithm in Case 2. The mean value of each step is about 0.07s.

To sum up, CAVs can form different coalitions to realize the collaborative decision making at the multi-lane merging conflict zone. The driving styles have significant effect on the coalition formation of CAVs. The basic rule of coalition formation is described in Definition 6.2. In the same coalition, CAVs can cooperate each other to maximize the interest of the coalition, which is beneficial to the performance improvement of the entire traffic system. The relationship between different coalitions is the competitive relationship, which is beneficial to the personalized driving. Namely, the proposed coalitional game-theoretic approach can consider the individual interest of each CAV and the whole interest of the entire traffic system in the decision-making process.

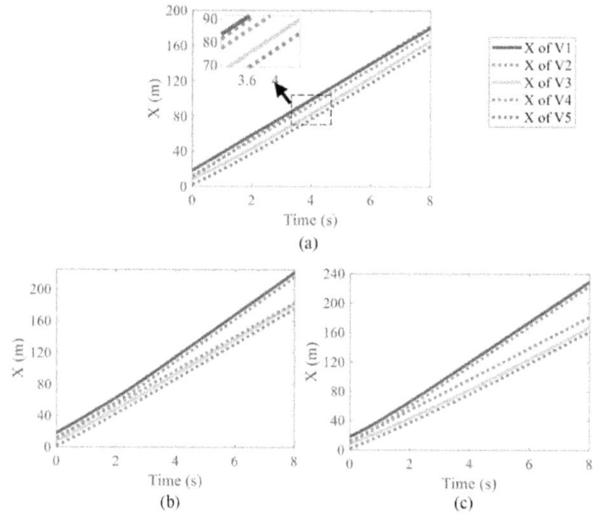

FIGURE 6.12: Longitudinal paths of CAVs under the three scenarios in Case 2: (a) Scenario A; (b) Scenario B; (c) Scenario C.

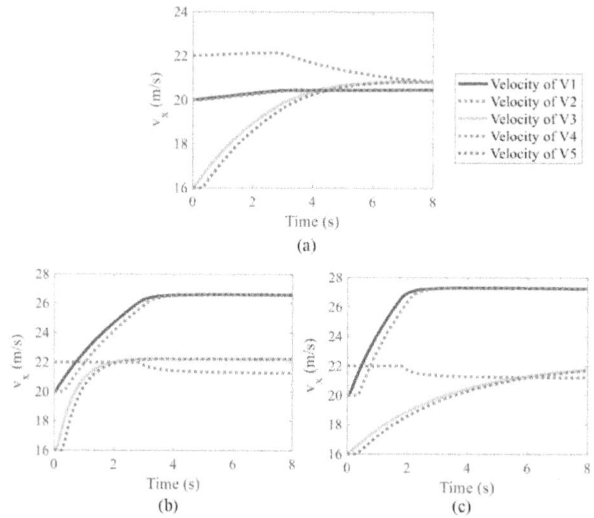

FIGURE 6.13: Velocities of CAVs under the three scenarios in Case 2: (a) Scenario A; (b) Scenario B; (c) Scenario C.

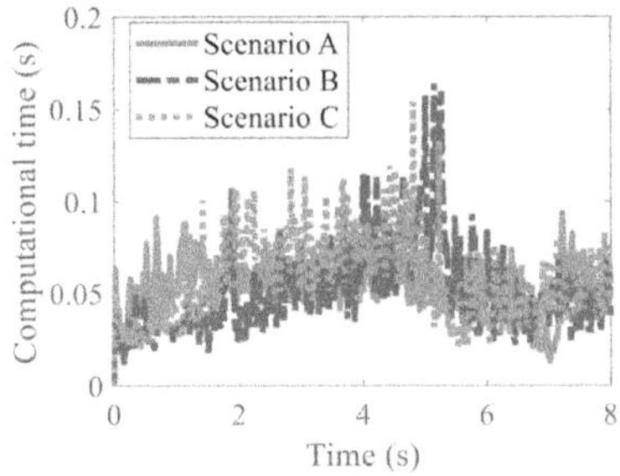

FIGURE 6.14: Computational time in Case 2.

6.2.5 Summary

To resolve the driving conflicts of CAVs at the multi-lane merging zone, a collaborative decision-making framework is proposed based on the coalitional game-theoretic approach. Firstly, a model-based motion prediction algorithm is designed to advance the decision-making safety. Comprehensively considering driving safety, ride comfort and travel efficiency, an integrated decision-making cost function is designed. Different driving styles are considered for human-like driving, reflecting different preferences on the three driving performances. Four typical coalition types are proposed, i.e., single player coalition, multi-player coalition, grand coalition and sub-coalition. CAVs can form different coalitions to pass the multi-lane merging zone. Based on the constructed decision-making cost function and multiple constraints, the coalitional game-theoretic approach is combined with MPC to resolve the driving conflicts of CAVs at the multi-lane merging zone. Finally, two test cases are designed and carried out to verify the proposed decision-making algorithm. The test results show that the coalitional game-theoretic approach can make safe decisions for CAVs at the multi-lane merging zone. It is worth mentioning that both competition and cooperation are considered in the coalitional game-theoretic approach, in favor of the personalized driving of each CAV and the performance improvement of the entire traffic system.

6.3 Cooperative Decision Making of CAVs at Unsignalized Roundabouts

The decision-making issue of AVs at unsignalized intersections has been described in Chapter 5. For cooperative decision making, all AVs described at unsignalized intersections are assumed to be CAVs in this chapter. The modeling and payoff function design work will not be introduced repeatedly. Based on the decision-making model constructed in Chapter 5, the cooperative game theoretic approach is used to address the driving conflict and interaction between CAVs at unsignalized intersections.

6.3.1 Decision Making with the Cooperative Game Theory

As Figure 6.15 shows, the grand coalition game approach is utilized to address the decision-making issue of CAVs at the unsignalized roundabout. In the grand coalition game, all the players form a large group that aims to maximize the payoff of the grand coalition. The grand coalition game is a typical cooperative game.

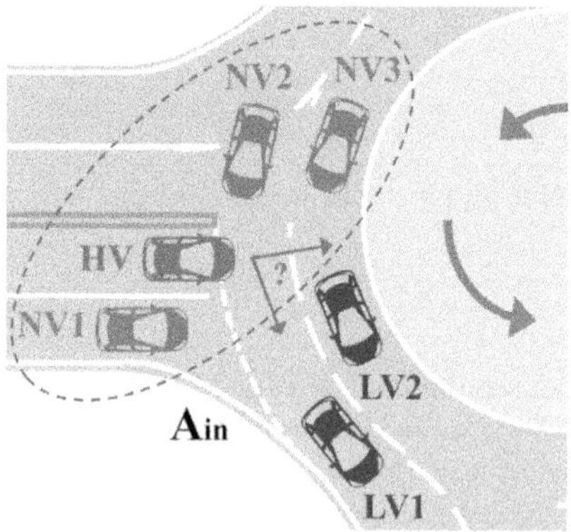

FIGURE 6.15: Decision making of CAVs at the roundabout with the cooperative game approach.

Based on the payoff function constructed in Chapter 5, the grand coalition game approach for decision making can be described as follows, which aims to maximize the payoff of the whole grand coalition.

$$\begin{aligned}(\hat{u}^{HV*}, \hat{u}^{NV1*}, \hat{u}^{NV2*}, \hat{u}^{NV3*}) &= \arg\max(\omega^{HV} P^{HV} \\ &+\omega^{NV1} P^{NV1} + \omega^{NV2} P^{NV2} + \omega^{NV3} P^{NV3})\end{aligned} \tag{6.54}$$

s.t. $\Xi^{HV} \leq 0$, $\Xi^{NV1} \leq 0$, $\Xi^{NV2} \leq 0$, $\Xi^{NV3} \leq 0$.
where ω^{HV}, ω^{NV1}, ω^{NV2} and ω^{NV3} denote the allocation coefficients of HV, NV1, NV2 and NV3, respectively. $\omega^{HV} + \omega^{NV1} + \omega^{NV2} + \omega^{NV3} = 1$.

Combined with the motion prediction of CAVs, the MPC approach is applied to the decision-making optimization.

At the time step k, the predicted payoff function sequence for CAVi is derived as follows.

$$P^i(k+1|k), P^i(k+2|k), \cdots, P^i(k+N_p|k) \tag{6.55}$$

To realize prediction-based decision making with MPC optimization, the following cost function is constructed.

$$J^i = 1/(P^i + \varepsilon) \tag{6.56}$$

where the coefficient $\varepsilon \to 0, \varepsilon > 0$.

The decision-making sequence for CAVi is expressed by

$$\hat{u}^i(k|k), \hat{u}^i(k+1|k), \cdots, \hat{u}^i(k+N_c-1|k) \tag{6.57}$$

where $\hat{u}^i(q|k) = [\Delta a_x^i(q|k), \Delta\delta_f^i(q|k), \alpha^i(q|k), \beta^i(q|k)]^T$, $q = k, k+1, \cdots, k+N_c-1$.

Then, the performance function of CAVi for decision making is derived by

$$\Pi^i = \sum_{p=k+1}^{k+N_p} ||J^i(p|k)||_Q^2 + \sum_{q=k}^{k+N_c-1} ||\hat{u}^i(q|k)||_R^2 \tag{6.58}$$

where Q and R denote the weighting matrices.

Finally, the grand coalition game approach for cooperative decision making is realized with MPC optimization.

$$\begin{aligned}(\hat{\mathbf{u}}^{HV*}, \hat{\mathbf{u}}^{NV1*}, \hat{\mathbf{u}}^{NV2*}, \cdots, \hat{\mathbf{u}}^{NVj*}) &= \arg\min[\omega^{HV}\Pi^{HV} \\ &+\omega^{NV1}\Pi^{NV1} + \omega^{NV2}\Pi^{NV2} + \cdots + \omega^{NVj}\Pi^{NVj}] \\ \text{s.t.} \quad \Xi^{HV} \leq 0, \Xi^{NV1} \leq 0, \Xi^{NV2} &\leq 0, \cdots, \Xi^{NVj} \leq 0.\end{aligned} \tag{6.59}$$

where the weighting coefficients follow that $\omega^{HV} + \omega^{NV1} + \omega^{NV2} + \cdots + \omega^{NVj} = 1$.

6.3.2 Testing Results and Analysis

To evaluate the performance of the cooperative game theoretic decision-making algorithm at the unsignalized roundabout, three test cases are designed and carried out. To conduct the comparative study with the noncooperative game theoretic approach, the test case is the same to that in Chapter 5.

6.3.2.1 Testing Case 1

In this case, it aims to test the algorithm performance that resolves the driving conflicts at the zone of entering the roundabout. HV on the inside lane of the main road $M1$ wants to enter the round road RR from the entrance A_{in} and exit from B_{out} to the main road $\hat{M}2$. At the entering stage, HV must interact with three NVs, i.e., NV1 on the outside lane of the main road $M1$, NV2 on the outside lane of the round road RR and NV3 on the inside lane of the round road RR, and then make the optimal decision.

Moreover, three typical scenarios are designed in this case to evaluate the effect of personalized driving on decision making. In Scenario A, the driving styles of HV, NV1, NV2 and NV3 are set as normal, conservative, normal and normal, respectively. In Scenario B, the driving styles of the four CAVs are set as normal, aggressive, normal and normal, respectively. In Scenario C, the driving styles of the four CAVs are set as aggressive, aggressive, normal and normal, respectively. In Case 1, the initial positions of HV, NV1, NV2, NV3, LV1 and LV2 are set as (-25, -2.45), (-28, -6.08), (-16, 10.25), (-10, 11.18), (-16, -10.25) and (-14, -5.38), respectively. In addition, the initial velocities of HV, NV1, NV2, NV3, LV1 and LV2 are set as 5.5 m/s, 4 m/s, 5 m/s, 4 m/s, 8 m/s, 8 m/s, respectively. In this case, the performances of the grand coalition game (GC) approach is tested and verified. The test results are presented in Figures 6.16 to 6.21.

The decision-making results in this case are illustrated in Figure 6.16. We can see the similar conclusions to the SG approach used in Chapter 5. Different scenarios show various decision-making results due to different driving styles of CAVs. In Scenario A, NV1 is set as the conservative driving styles, resulting in the slowing

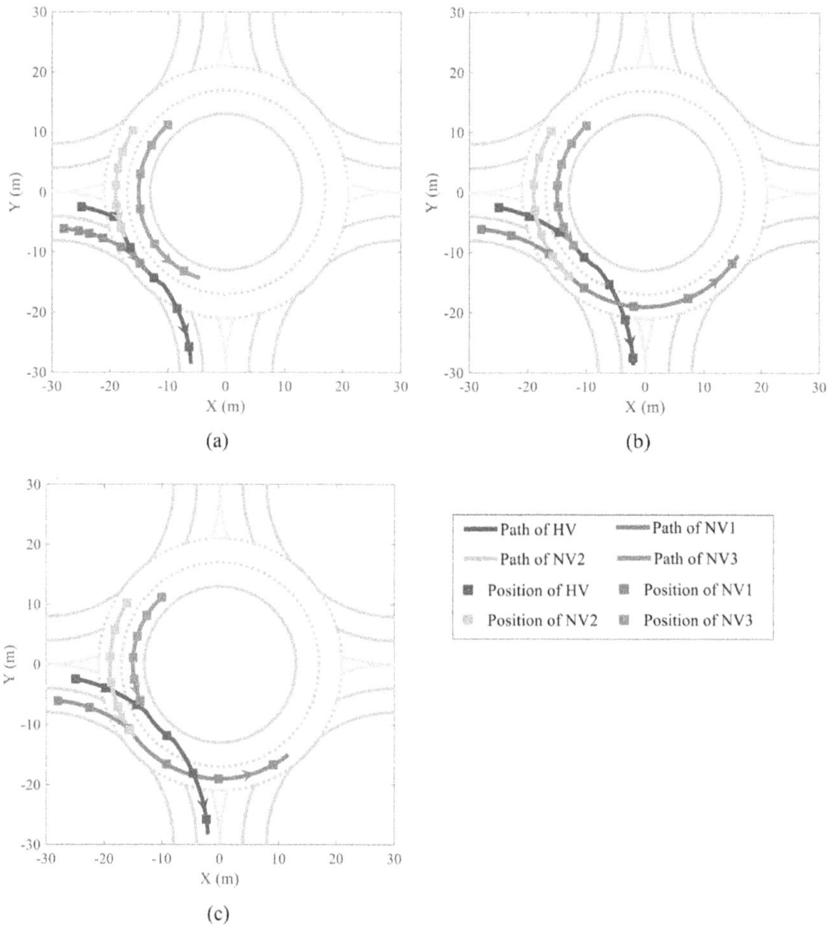

FIGURE 6.16: Decision-making results of CAVs in Case 1: (a) Scenario A; (b) Scenario B; (c) Scenario C.

down and giving ways to HV. HV merged into the outside lane of the round road *RR*. As a result, NV2 had to decelerate to keep a safe distance from HV. Since there is no driving conflict for NV3, the motion of NV3 is normal. Compared with the SG approach, it can be found from Figure 6.17 that the speed of NV1 is larger in the GC approach the and there is no sudden deceleration for NV1, indicating that the GC approach can realize collaborative decision making for all CAVs. The travel efficiency of the entire traffic system is increased, which is the great advantage of the cooperative game theoretic approach. In Scenario B, all CAVs are the normal driving style. HV merged into the inside lane of the round road *RR*. As a result, NV3 had to slow down and give ways for HV. Moreover, NV2 decelerated to keep a safe distance from NV1. In Scenario C, it is an extreme condition that both HV and

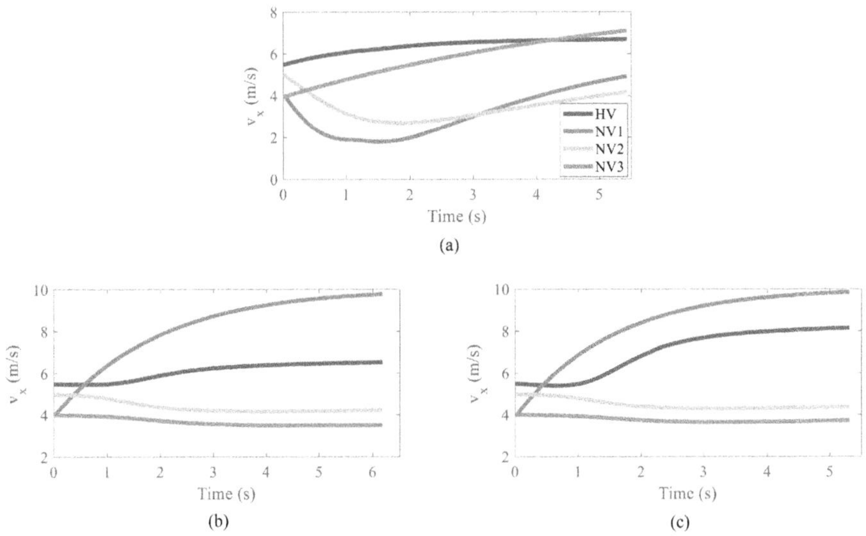

FIGURE 6.17: Velocities of CAVs in Case 1: (a) Scenario A; (b) Scenario B; (c) Scenario C.

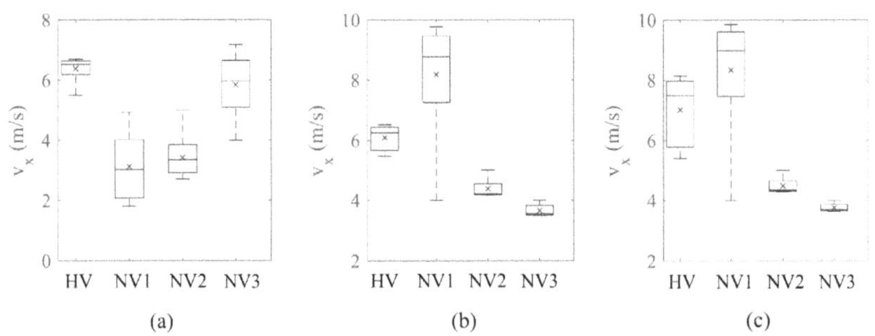

FIGURE 6.18: Box plots of velocities in Case 1: (a) Scenario A; (b) Scenario B; (c) Scenario C.

NV1 are the aggressive driving style. Figure 6.17(c) shows that both HV and NV1 have larger travel velocity than NV2 and NV3, which indicates that the aggressive driving style pursues higher travel efficiency. Besides, we can find from Figure 6.16(c) that HV merged into the inside lane of the round road *RR* and NV1 merged into the outside lane of the round road *RR*. As a result, NV2 and NV3 had to decelerate to give ways for NV1 and HV, respectively. Compared with the SG approach, it can be found from Figure 6.17 that the velocities of NV1 and HV are smaller in the GC approach. Namely, the effect of driving style on decision making is weakened in the GC approach. The GC approach cares more about the interest of the entire traffic system. However, the SG approach gives more priority on the individual interest.

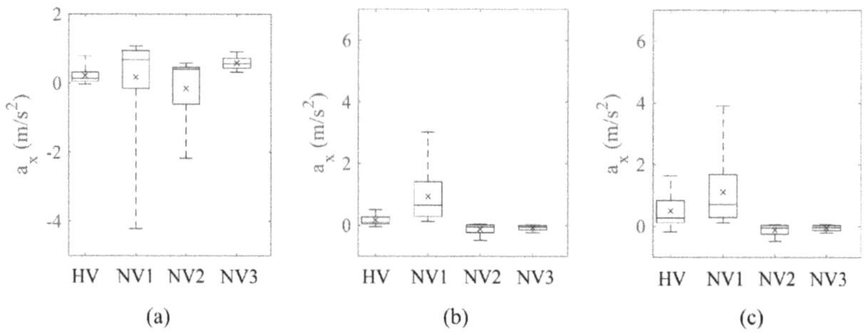

FIGURE 6.19: Box plots of longitudinal accelerations in Case 1: (a) Scenario A; (b) Scenario B; (c) Scenario C.

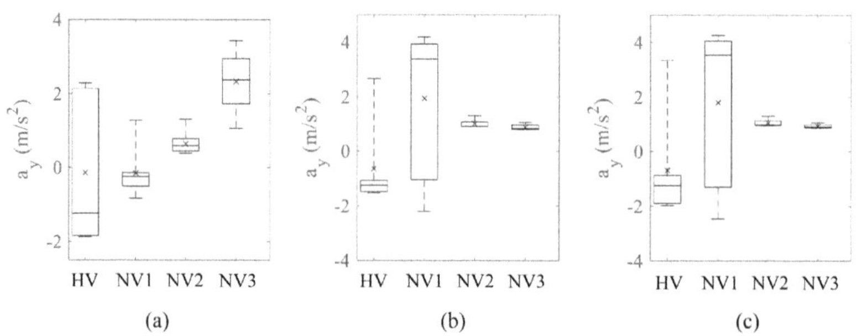

FIGURE 6.20: Box plots of lateral accelerations in Case 1: (a) Scenario A; (b) Scenario B; (c) Scenario C.

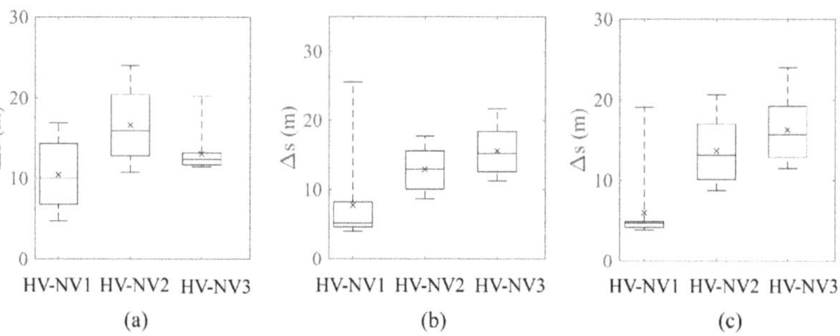

FIGURE 6.21: Box plots of relative distances between HV and other NVs in Case 1: (a) Scenario A; (b) Scenario B; (c) Scenario C.

Besides, the distributions of longitudinal and lateral accelerations are illustrated in Figures 6.19 and 6.20, respectively. We can find that the aggressive driving style shows larger acceleration distribution than the normal and conservative driving styles, which means that the aggressive driving style cares more about travel efficiency rather than ride comfort. However, the conservative driving style has the opposite preference. Moreover, we can see from Figure 6.21 that the aggressive driving style usually shows the smaller safe distance, especially for the two aggressive drivers. As a result, the driving safety is worsened remarkably. The conservative driving style shows larger safe distance, indicating that the conservative driving style gives more priority on driving safety. The normal driving style is between the above two driving styles, finding a good balance between different driving performances.

6.3.2.2 Testing Case 2

In Case 2, it aims to address the driving conflicts and decision-making issue that CAVs pass and exit the roundabout. HV on the outside lane of the round road RR is ready to move out from D_{out} to the main road $\hat{M}4$. In this process, HV has to resolve the merging conflict from NV1 on the main road $M3$. HV has three choices, i.e., slowing down and giving ways to NV1, speeding up and fighting for the right of way, or changing lanes to the inside lane of the round road RR. If HV chooses the last one, it must interact with NV2. The merging conflict from NV1 is transformed into a lane-change conflict.

In this case, to simulate the human-like driving and decision making, different driving styles are set for CAVs. In Scenario A, the driving styles of HV, NV1 and NV2 are set as conservative, normal and normal, respectively. In Scenario B, the driving styles of the three CAVs are all normal. In Scenario C, the driving styles of the three AVs are set as aggressive, normal and normal, respectively. Besides, the initial positions of HV, NV1, NV2, LV1 and LV2 are set as (15, -11.66), (25, 2.45), (8, -12.68), (15, 11.66) and (13, 7.48), respectively. The initial velocities of HV, NV1, NV2, LV1 and LV2 are set as 5.5 m/s, 5 m/s, 4 m/s, 8 m/s, 5 m/s, respectively. The testing results are displayed in Figures 6.22 to 6.27.

The decision-making results in Case 2 are displayed in Figure 6.22. Different driving styles of CAVs lead to various test results in the three Scenarios. In Scenario A, HV made the decision slowing down and giving ways for NV1 due to the conservative driving style of HV. Therefore, we can see a sudden deceleration of HV in Figure 6.23(a). With the deceleration behavior, the driving safety of HV is advanced. In Scenario B, HV chose lane-change to resolve the merging conflict of NV1. As a result, NV2 had to decelerate and give ways for HV. In Scenario C, HV had a double-lane change behavior to resolve the merging conflict of NV1. From Figure 6.23(c), we can see that both NV1 and NV2 have obvious deceleration behaviors. However, the travel velocity of HV is much larger, indicating that the aggressive driving style pursues higher travel efficiency. Compared with the test results in the SG approach, the travel velocity of HV is decreased and the travel velocity of NV1 and NV2 are increased in the GC approach. We can conclude that the personalized driving is weakened in the GC approach, but the system efficiency is increased.

Moreover, the distributions of longitudinal and lateral accelerations are displayed in Figures 6.25 and 6.26, respectively. We can find that the aggressive driving style has larger acceleration to improve the travel efficiency. However, the large acceleration will worsen the ride comfort. The conservative driving style shows smaller acceleration to advance the ride comfort. Besides, Figure 6.27 shows the relative distances between CAVs. We can find that the conservative driving style shows larger safe distance to enhance the driving safety. However, the aggressive driving style shows the opposite result.

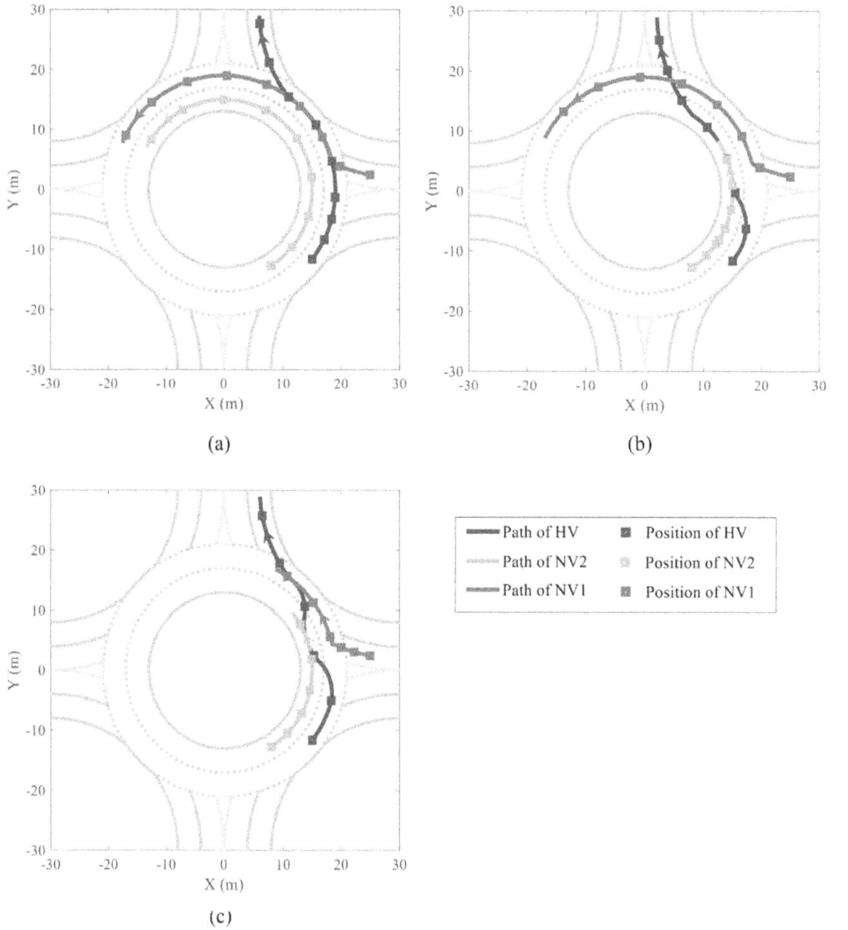

FIGURE 6.22: Decision-making results of CAVs in Case 2: (a) Scenario A; (b) Scenario B; (c) Scenario C.

6.3.2.3 Testing Case 3

This case considers five CAVs at the unsignalized roundabout. All the three stages are included in this case, and the travel efficiency of the entire traffic system is evaluated. At the initial time, HV on the outside lane of the main road $M1$ is ready to enter the round road RR from the entrance A_{in} and exit from A_{out} to the main road $\hat{M}1$, simulating a U-turn scenario. Besides HV, four NVs are considered at the unsignalized roundabout. NV1 on the inner lane of the round road RR starts to exit from C_{out} to the main road $\hat{M}3$. NV2 on the external lane of the main road $M2$ is ready to enter the round road RR and exit from C_{out} to the main road $\hat{M}3$. NV3 on the inner lane of the main road $M3$ starts to enter the round road RR and

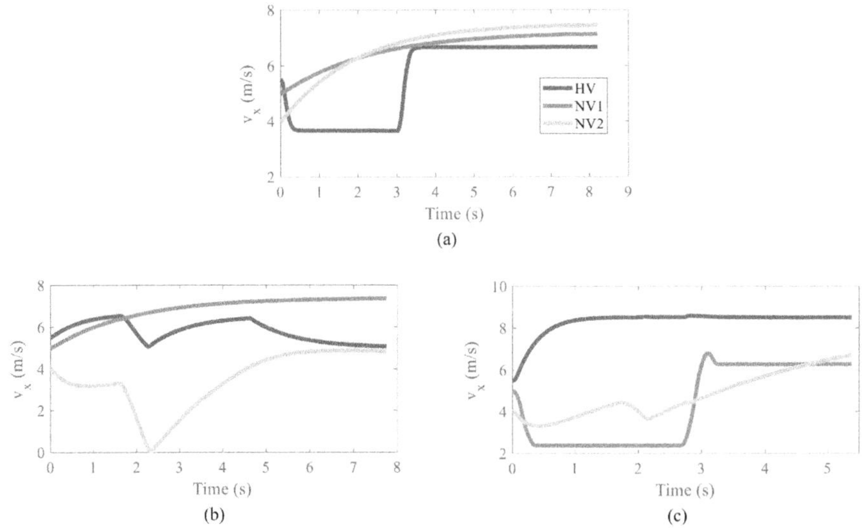

FIGURE 6.23: Velocities of CAVs in Case 2: (a) Scenario A; (b) Scenario B; (c) Scenario C.

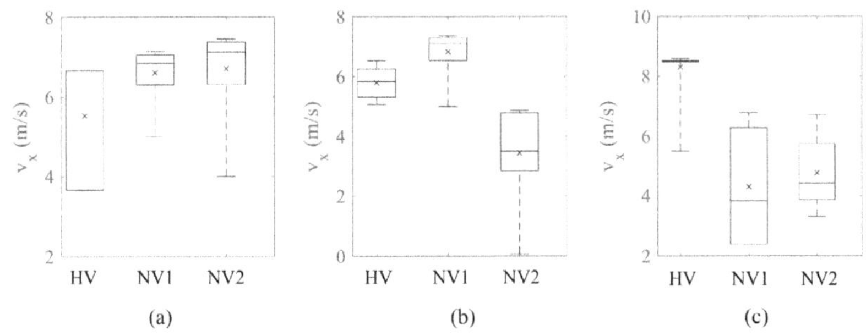

FIGURE 6.24: Box plots of velocities in Case 2: (a) Scenario A; (b) Scenario B; (c) Scenario C.

exit from A_{out} to the main road $\hat{M}1$. NV4 on the outer lane of the main road $M4$ wants to enter the round road RR from the entrance D_{in} and exit from D_{out} to the main road $\hat{M}4$.

Besides, to study the effect of personalized driving on the travel efficiency of the entire traffic system, different driving styles are considered for CAVs. The driving style of HV is set as aggressive and the driving styles of four NVs are set as normal. The initial positions of HV, NV1, NV2, NV3 and NV4 are set as (-19, -8.68), (-9, -12), (6, -35), (50, 2) and (-6, 88), respectively. The initial velocities of HV, NV1, NV2, NV3 and NV4 are set as 5.5 m/s, 5 m/s, 5 m/s, 4 m/s, 4 m/s, respectively. The testing results are displayed in Figures 6.28 to 6.30.

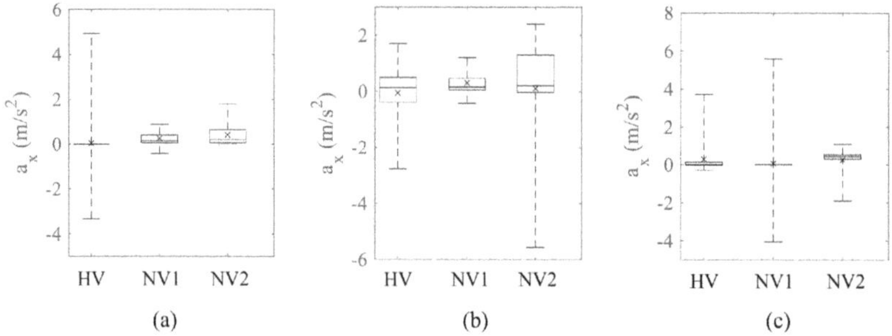

FIGURE 6.25: Box plots of longitudinal accelerations in Case 2: (a) Scenario A; (b) Scenario B; (c) Scenario C.

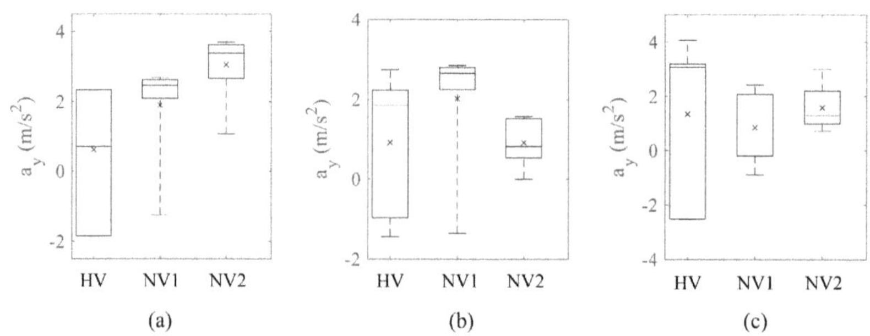

FIGURE 6.26: Box plots of lateral accelerations in Case 2: (a) Scenario A; (b) Scenario B; (c) Scenario C.

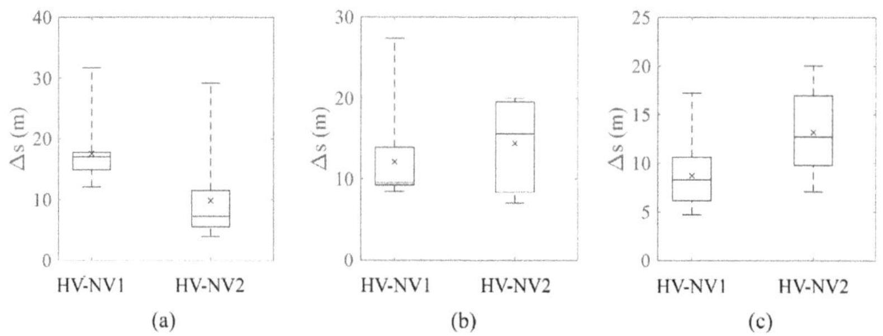

FIGURE 6.27: Box plots of relative distances between HV and other NVs in Case 2: (a) Scenario A; (b) Scenario B; (c) Scenario C.

The decision-making results of CAVs are illustrated in Figure 6.28. We can find that HV conducted two times lane-change to resolve the driving conflicts from other NVs. The first driving conflict for HV comes from the merging behavior of NV2. HV made the decision changing lanes to the inside lane of the round road. To avoid

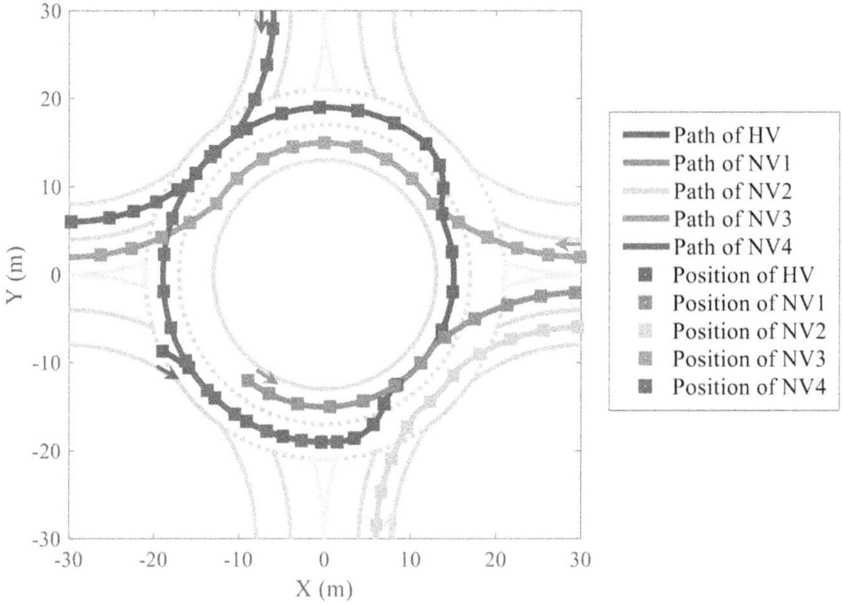

FIGURE 6.28: Decision-making results of CAVs in Case 3

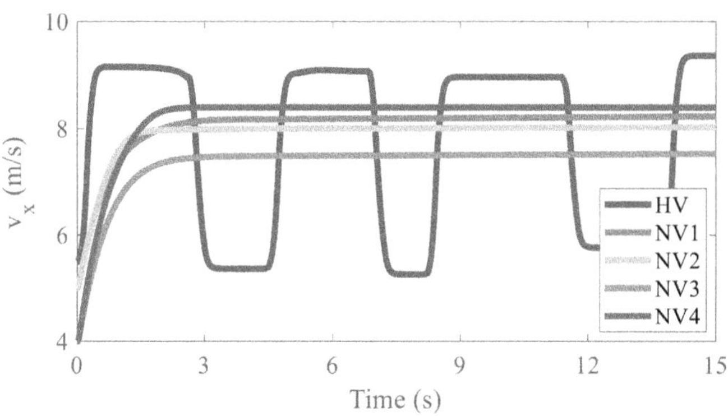

FIGURE 6.29: Velocities of CAVs in Case 3.

the collision with NV1, HV slowed down. We can see an obvious deceleration of HV in Figure 6.29. Furthermore, to address the merging conflict of NV3, HV changed lanes again to the outsides lane of the round road. Finally, to resolve the driving conflict from NV4, HV made the decision slowing down and giving ways for NV4.

TABLE 6.4: Travel Efficiency Analysis in Case 3

Testing Results	HV	NV1	NV2	NV3	NV4
Velocity Max / (m/s)	9.36	8.23	8.03	7.53	8.40
Velocity RMS / (m/s)	7.85	8.05	7.91	7.36	8.22
System Velocity RMS / (m/s)			7.88		

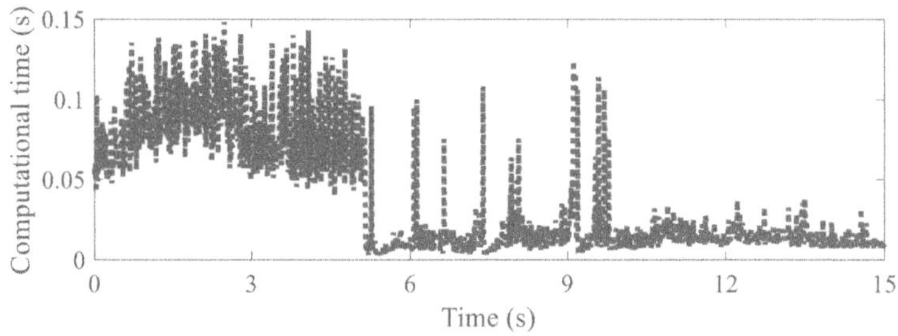

FIGURE 6.30: Computational time of the proposed algorithm.

From Figure 6.29, we can see the frequent acceleration and deceleration behaviors of HV. Table 6.4 shows the travel efficiency analysis of all CAVs and the traffic system. We can find that HV has the largest passing velocity, indicating that the aggressive style pursues higher travel efficiency. Compared with the analysis results of the SG approach in Chapter 5, we can find that the travel velocity of HV is decreased, and the system velocity is increased. It can be concluded that the SG approach is in favor of the personalized driving. However, the GC approach is beneficial to the collaborative decision making of CAVs, increasing the system efficiency.

Moreover, the computational time of the GC approach in Case 3 is displayed in Figure 6.30. The mean value of computational time for each iteration of the GC approach is 0.038s, which is smaller than that of the SG approach. It can be explained that the interaction behavior of CAVs in the GC approach is smaller than that in the SG approach, which is an advantage of the GC approach for the collaborative decision making of CAVs.

6.3.2.4 Discussion

Based on the test results of the driving conflict resolution at the unsignalized roundabout in Chapter 5 and this chapter, we can conclude that both the SG approach and the GC approach can realize human-like decision making for autonomous driving and meanwhile resolve all kinds of driving conflicts. Different driving styles are considered for AVs. The aggressive driving style gives more priority on travel efficiency, and the conservative driving style prefers higher ride comfort and driving safety. The two game theoretic decision-making algorithms can be used for different objectives. The SG approach cares more about the individual interest, which is beneficial to the personalized driving. The GC approach cares more about the system interest, which is beneficial to the collaborative decision making.

6.3.3 Summary

In this section, a cooperative game theoretic approach is applied to the collaborative decision-making issue of CAVs at the unsignalized roundabout. The modeling work and decision-making payoff function are the same to that in Chapter 5. Only the decision-making algorithm is different, i.e., the noncooperative game theoretic approach in Chapter 5 and the cooperative game theoretic approach in Chapter 6. The same driving scenario and test case are designed and carried out. From the test results, it can be concluded that both the two game theoretic approaches can make feasible and reasonable decisions for CAVs at the unsignalized roundabout. The driving conflicts at the complex roundabout zone can be resolved. The noncooperative game theoretic approach is in favor of the personalized driving. However, the cooperative game theoretic approach is beneficial to the efficiency improvement of the entire traffic system.

6.4 Conclusion

This chapter aims to resolve the driving conflicts of CAVs in complex traffic scenarios, i.e., highway merging and unsignalized roundabout. The cooperative game theoretic approaches are applied to the collaborative decision making of CAVs. To address the multi-lane merging issue of CAVs, a cooperative game theoretic decision-making framework is designed. The motion prediction model is proposed in favor of the performance advancement of the decision-making algorithm. In the decision-making cost function design, various driving performances are considered including safety, comfort and efficiency. Based on the decision-making cost function, the coalitional game approach is used to address the driving conflicts of CAVs at unsignalized roundabouts. Four typical coalition types are proposed for CAVs to conduct the collaborative decision making. Finally, different scenarios and driving characteristics of CAVs are tested. The test results indicate that the developed coalitional game approach is able to make feasible and reasonable decisions for CAVs at the multi-lane merging zone, and the resultant different coalition formations are adaptive to various driving scenarios.

To resolve the driving conflicts of CAVs at unsignalized roundabouts, a cooperative game theoretic decision-making framework is designed with the grand coalitional game. The comparative study with the noncooperative game theoretic approach in Chapter 5 is conducted. The test results indicate that both the noncooperative game theoretic approach and the cooperative game theoretic approach can address the driving conflicts of CAVs at unsignalized roundabouts with safe decision making. The noncooperative game theoretic approach cares more about personalized driving, while the cooperative game theoretic approach is beneficial to the performance improvement of the entire traffic system, which shows the difference between two approaches.

Chapter 7

Conclusion, Discussion and Prospects

In this book, the human-like decision making and control technology is studied for autonomous driving with game theoretic approaches. The feature identification and modeling of human-like driving behaviors are conducted using the real world driving dataset. Based on this, AVs are given different driving styles to realize human-like driving, reflecting different driving preferences on safety, comfort and efficiency. Different game theoretic approaches are applied to the decision-making issues in different driving scenarios, including lane-change, merging on highways and unsignalized roundabout. To sum up, the proposed human-like decision-making framework can support the personalized driving for AVs. Besides, the collaborative decision making between CAVs can address the driving conflicts in complex traffic scenarios, advancing the driving safety and travel efficiency.

In the following sections, some summarizations, challenges and prospects regarding the human-like driving of AVs are presented.

7.1 Human-Like Modeling for AVs

In this book, the work of human-like modeling for AVs is divided into two parts. The first part is the human-like driving feature identification and representation. The commonest method is the driving style division according to the real world driving dataset. Referring the driving styles of human drivers, AVs can simulate the driving behaviors of human drivers. The challenge is that the human-like driving styles are usually limited. For instance, the commonest classification is three driving styles, i.e., aggressive, normal and conservative. Although human drivers are divided into different driving styles, it is a statistical result. The driving style of one human driver is not always the fixed one, which is affected by some factors, e.g., physical and psychological state, weather, road conditions, etc. In existing studies, the human-like driving styles of AVs are usually fixed, which cannot represent the changeable driving styles like human drivers. To realize changeable driving styles, the concept of driving aggressiveness is proposed. In Chapter 2, we applied the longitudinal

DOI: 10.1201/9781003287087-7

velocity and yaw rate to the driving aggressiveness analysis. With the application of the fuzzy inference approach, the driving aggressiveness is obtained, which is a continuous variable from zero to one. Then, the human-like driving feature can be represented continuously. In this book, only the longitudinal velocity and yaw rate are used to reflect the driving aggressiveness. For future work, more performance parameters can be used for more accurate description of the driving aggressiveness.

After finishing the work of human-like driving feature identification and representation, it comes to the second part, i.e., the modeling of human-like driving behaviors. According to the driving styles, different weights are allocated to different driving performances, i.e., driving safety, ride comfort, and travel efficiency. As a result, AVs can demonstrate human-like driving behaviors. In this study, the weight allocation is fixed for each driving style. With the application of the driving aggressiveness, the weighting parameters can be time-varying, which is more reasonable and human-like. To realize human-like path planning, different driving styles are reflected by different APF distributions in this book. In the human-like control model, the driver model is used, and different parameter settings can simulate different driving styles. In the future work, brain-like models can be used for the human-like control of AVs.

7.2 Human-Like Decision-Making Algorithm

In the human-like decision-making algorithm, the decision-making cost function design is very critical, which involves different performance indexes and multiple decision-making variables. As mentioned in above section, the performance indexes include driving safety, ride comfort, and travel efficiency. Driving safety is reflected by the relative distance and velocity of two vehicles. Ride comfort is associated with the acceleration and jerk, and travel efficiency is related to the vehicle velocity. The decision-making variables include the behaviour variable, e.g., lane-change and lane-keeping, and the action variable, e.g., steering angle and acceleration. For different driving scenarios, the behavior variables may be different. For instance, at the roundabout, there exist the behavior variables merging into the inside lane of round road, lane keeping, and merging into the outside lane of round road. However, the behavior variables for highways are lane-change and lane-keeping. Based on above analysis, it can be found that we must reconstruct the decision-making cost function for different driving scenarios. In the future work, the comprehensive decision-making cost function will be constructed to deal with different driving scenarios. Meanwhile, all kinds of decision-making variables will be considered, improving the generalization of the decision-making algorithm.

To realize human-like interaction and decision making, different game theoretic approaches are utilized in this book, including Nash equilibrium and Stackelberg game. In the lane-change scenario, both two approaches can make human-like decisions under different driving styles. Compared with the Nash equilibrium approach, the Stackelberg game approach shows better performance in terms of driving safety, ride comfort and travel efficiency. In the lane-change interaction process, the lane-change vehicle is the leader that make the decision firstly, and the obstacle vehicle is the follower that make the decision later, which follows the concept of Stackelberg game. Therefore, the Stackelberg game approach is more human-like in the lane-change scenario. In this study, only few players are considered in the interaction process. To deal with the interaction and decision making in the high density and dynamic traffic flow environment, e.g., traffic jam, the multi-leader-follower Stackelberg game approach will be utilized in the future.

7.3 Cooperative Decision Making Considering Personalized Driving

To resolve the driving conflicts in complex traffic scenarios, e.g., merging on highways and unsignalized roundabout, a cooperative decision-making framework is designed for CAVs using a coalitional game approach. The proposed coalitional game based method is able to make reasonable decisions and adapt to different driving characteristics for CAVs. It guarantees the safety and efficiency of CAVs at the complex dynamic traffic condition, and simultaneously accommodates the objectives of individual vehicles. Nonetheless, there are still some challenges. With the increase of the system complexity, e.g., more CAVs in the decision-making model, a larger amount of computation resources and a more powerful execution capability for decision making are required. Therefore, further improvement of the computation efficiency of the algorithm is a critical direction of the future work. We consider this issue from two aspects. On one hand, we can optimize the algorithm and further reduce the algorithm complexity, e.g., designing the event trigger mechanism to reduce the unnecessary interaction process. On the other hand, we will design better algorithm solver to the algorithm efficiency. If the real-time performance can be guaranteed, the proposed decision-making algorithm will be applied to more complex traffic scenarios.

Additionally, motion prediction is in favor of the performance improvement of the decision-making algorithm, especially for driving safety. Therefore, the motion prediction module is very critical for the decision-making framework. In this study, the model-based motion prediction algorithm is designed. The prediction accuracy is acceptable in the short-term motion prediction. However, the prediction accuracy is greatly reduced in the long-term motion prediction. Especially for the extreme condition, long-term motion prediction can remarkably improve the decision-making safety. To this end, we will apply the learning-based approach, e.g., LSTM algorithm, to the long-term motion prediction in the future work.

Bibliography

[1] Yasir Ali, Zuduo Zheng, Md Mazharul Haque, and Meng Wang. A game theory-based approach for modelling mandatory lane-changing behaviour in a connected environment. *Transportation Research Part C: Emerging Technologies*, 106:220–242, 2019.

[2] Ahmad Aljaafreh, Nabeel Alshabatat, and Munaf S Najim Al-Din. Driving style recognition using fuzzy logic. In *2012 IEEE International Conference on Vehicular Electronics and Safety (ICVES 2012)*, pages 460–463. IEEE, 2012.

[3] Sharaf Alkheder, Fahad Al-Rukaibi, and Asma Al-Faresi. Driver behavior at kuwait roundabouts and its performance evaluation. *IATSS research*, 44(4):272–284, 2020.

[4] Noor Hafizah Amer, Hairi Zamzuri, Khisbullah Hudha, and Zulkiffli Abdul Kadir. Modelling and control strategies in path tracking control for autonomous ground vehicles: a review of state of the art and challenges. *Journal of Intelligent & Robotic Systems*, 86(2):225–254, 2017.

[5] Adel Ammar, Hachemi Bennaceur, Imen Châari, Anis Koubâa, and Maram Alajlan. Relaxed dijkstra and a* with linear complexity for robot path planning problems in large-scale grid environments. *Soft Computing*, 20(10):4149–4171, 2016.

[6] Aliasghar Arab, Kaiyan Yu, Jingang Yi, and Dezhen Song. Motion planning for aggressive autonomous vehicle maneuvers. In *2016 IEEE International Conference on Automation Science and Engineering (CASE)*, pages 221–226. IEEE, 2016.

[7] Antonio Artuñedo, Jorge Godoy, and Jorge Villagra. Smooth path planning for urban autonomous driving using openstreetmaps. In *2017 IEEE Intelligent Vehicles Symposium (IV)*, pages 837–842. IEEE, 2017.

[8] Mansour Ataei, Amir Khajepour, and Soo Jeon. Model predictive control for integrated lateral stability, traction/braking control, and rollover prevention of electric vehicles. *Vehicle System Dynamics*, 58(1):49–73, 2020.

[9] Haoyu Bai, Shaojun Cai, Nan Ye, David Hsu, and Wee Sun Lee. Intention-aware online pomdp planning for autonomous driving in a crowd. In *2015 IEEE International Conference on Robotics and Automation (ICRA)*, pages 454–460. IEEE, 2015.

[10] Soohyuk Bang and Soyoung Ahn. Control of connected and autonomous vehicles with cut-in movement using spring mass damper system. *Transportation Research Record*, 2672(20):133–143, 2018.

[11] Craig Earl Beal and J Christian Gerdes. Model predictive control for vehicle stabilization at the limits of handling. *IEEE Transactions on Control Systems Technology*, 21(4):1258–1269, 2012.

[12] Mohammad Mahdi Bejani and Mehdi Ghatee. A context aware system for driving style evaluation by an ensemble learning on smartphone sensors data. *Transportation Research Part C: Emerging Technologies*, 89:303–320, 2018.

[13] Tomas Berglund, Andrej Brodnik, Håkan Jonsson, Mats Staffanson, and Inge Soderkvist. Planning smooth and obstacle-avoiding b-spline paths for autonomous mining vehicles. *IEEE Transactions on Automation Science and Engineering*, 7(1):167–172, 2009.

[14] Youssef Bichiou and Hesham A Rakha. Developing an optimal intersection control system for automated connected vehicles. *IEEE Transactions on Intelligent Transportation Systems*, 20(5):1908–1916, 2018.

[15] Paul T Boggs and Jon W Tolle. Sequential quadratic programming. *Acta Numerica*, 4:1–51, 1995.

[16] Khac-Hoai Nam Bui and Jason J Jung. Cooperative game-theoretic approach to traffic flow optimization for multiple intersections. *Computers & Electrical Engineering*, 71:1012–1024, 2018.

[17] Khac-Hoai Nam Bui and Jason J Jung. Aco-based dynamic decision making for connected vehicles in iot system. *IEEE Transactions on Industrial Informatics*, 15(10):5648–5655, 2019.

[18] Görkem Büyükyildiz, Olivier Pion, Christoph Hildebrandt, Martin Sedlmayr, Roman Henze, and Ferit Küçükay. Identification of the driving style for the adaptation of assistance systems. *International Journal of Vehicle Autonomous Systems*, 13(3):244–260, 2017.

[19] Chun-Wei Chang, Chen Lv, Huaji Wang, Hong Wang, Dongpu Cao, Efstathios Velenis, and Fei-Yue Wang. Multi-point turn decision making framework for human-like automated driving. In *2017 IEEE 20th International Conference on Intelligent Transportation Systems (ITSC)*, pages 1–6. IEEE, 2017.

[20] Depeng Chen, Zhijun Chen, Yishi Zhang, Xu Qu, Mingyang Zhang, and Chaozhong Wu. Driving style recognition under connected circumstance using a supervised hierarchical bayesian model. *Journal of Advanced Transportation*, 2021, 2021.

[21] J. Chen, Z. Pan, H. Liang, and M. Tao. A multiple attribute-based decision making model for autonomous vehicle in urban environment. In *Intelligent Vehicles Symposium*, 2014.

[22] Jianyu Chen, Bodi Yuan, and Masayoshi Tomizuka. Model-free deep reinforcement learning for urban autonomous driving. In *2019 IEEE Intelligent Transportation Systems Conference (ITSC)*, pages 2765–2771. IEEE, 2019.

[23] Na Chen, Bart van Arem, Tom Alkim, and Meng Wang. A hierarchical model-based optimization control approach for cooperative merging by connected automated vehicles. *IEEE Transactions on Intelligent Transportation Systems*, 2020.

[24] Tingting Chen, Liehuang Wu, Fan Wu, and Sheng Zhong. Stimulating cooperation in vehicular ad hoc networks: A coalitional game theoretic approach. *IEEE Transactions on Vehicular Technology*, 60(2):566–579, 2010.

[25] Seungwon Choi, Nahyun Kweon, Chanuk Yang, Dongchan Kim, Hyukju Shon, Jaewoong Choi, and Kunsoo Huh. Dsa-gan: Driving style attention generative adversarial network for vehicle trajectory prediction. In *2021 IEEE International Intelligent Transportation Systems Conference (ITSC)*, pages 1515–1520. IEEE, 2021.

[26] K Chu, J Kim, K Jo, and Myoungho Sunwoo. Real-time path planning of autonomous vehicles for unstructured road navigation. *International Journal of Automotive Technology*, 16(4):653–668, 2015.

[27] Jorge Cordero, Jose Aguilar, Kristell Aguilar, Danilo Chávez, and Eduard Puerto. Recognition of the driving style in vehicle drivers. *Sensors*, 20(9):2597, 2020.

[28] Changhua Dai, Changfu Zong, and Guoying Chen. Path tracking control based on model predictive control with adaptive preview characteristics and speed-assisted constraint. *IEEE Access*, 8:184697–184709, 2020.

[29] Pierre De Beaucorps, Thomas Streubel, Anne Verroust-Blondet, Fawzi Nashashibi, Benazouz Bradai, and Paulo Resende. Decision-making for automated vehicles at intersections adapting human-like behavior. In *2017 IEEE Intelligent Vehicles Symposium (IV)*, pages 212–217. IEEE, 2017.

[30] Ezequiel Debada and Denis Gillet. Cooperative circulating behavior at single-lane roundabouts. In *2018 21st International Conference on Intelligent Transportation Systems (ITSC)*, pages 3306–3313. IEEE, 2018.

[31] Ezequiel Gonzale Debada and Denis Gillet. Virtual vehicle-based cooperative maneuver planning for connected automated vehicles at single-lane roundabouts. *IEEE Intelligent Transportation Systems Magazine*, 10(4):35–46, 2018.

[32] Inés del Campo, Estibalitz Asua, Victoria Martínez, Óscar Mata-Carballeira, and Javier Echanobe. Driving style recognition based on ride comfort using a hybrid machine learning algorithm. In *2018 21st International Conference on Intelligent Transportation Systems (ITSC)*, pages 3251–3258. IEEE, 2018.

[33] Zejian Deng, Duanfeng Chu, Chaozhong Wu, Shidong Liu, Chen Sun, Teng Liu, and Dongpu Cao. A probabilistic model for driving-style-recognition-enabled driver steering behaviors. *IEEE Transactions on Systems, Man, and Cybernetics: Systems*, 2020.

[34] Jishiyu Ding, Huei Peng, Yi Zhang, and Li Li. Penetration effect of connected and automated vehicles on cooperative on-ramp merging. *IET Intelligent Transport Systems*, 14(1):56–64, 2019.

[35] Nemanja Djuric, Vladan Radosavljevic, Henggang Cui, Thi Nguyen, Fang-Chieh Chou, Tsung-Han Lin, Nitin Singh, and Jeff Schneider. Uncertainty-aware short-term motion prediction of traffic actors for autonomous driving. In *Proceedings of the IEEE/CVF Winter Conference on Applications of Computer Vision*, pages 2095–2104, 2020.

[36] Yiqun Dong, Youmin Zhang, and Jianliang Ai. Experimental test of unmanned ground vehicle delivering goods using rrt path planning algorithm. *Unmanned Systems*, 5(01):45–57, 2017.

[37] Dominik Dörr, David Grabengiesser, and Frank Gauterin. Online driving style recognition using fuzzy logic. In *17th International IEEE Conference on Intelligent Transportation Systems (ITSC)*, pages 1021–1026. IEEE, 2014.

[38] Dominik Dörr, Konstantin D Pandl, and Frank Gauterin. Optimization of system parameters for an online driving style recognition. In *2016 IEEE 19th International Conference on Intelligent Transportation Systems (ITSC)*, pages 302–307. IEEE, 2016.

[39] Haibin Duan and Linzhi Huang. Imperialist competitive algorithm optimized artificial neural networks for ucav global path planning. *Neurocomputing*, 125:166–171, 2014.

[40] Jingliang Duan, Shengbo Eben Li, Yang Guan, Qi Sun, and Bo Cheng. Hierarchical reinforcement learning for self-driving decision-making without reliance on labelled driving data. *IET Intelligent Transport Systems*, 14(5):297–305, 2020.

[41] Azim Eskandarian, Chaoxian Wu, and Chuanyang Sun. Research advances and challenges of autonomous and connected ground vehicles. *IEEE Transactions on Intelligent Transportation Systems*, 2019.

[42] Filippo Fabiani and Sergio Grammatico. Multi-vehicle automated driving as a generalized mixed-integer potential game. *IEEE Transactions on Intelligent Transportation Systems*, 21(3):1064–1073, 2019.

[43] F Clara Fang and Hernan Castaneda. Computer simulation modeling of driver behavior at roundabouts. *International Journal of Intelligent Transportation Systems Research*, 16(1):66–77, 2018.

[44] Liangji Fang, Qinhong Jiang, Jianping Shi, and Bolei Zhou. Tpnet: Trajectory proposal network for motion prediction. In *Proceedings of the IEEE/CVF Conference on Computer Vision and Pattern Recognition*, pages 6797–6806, 2020.

[45] Dennis Fassbender, Benjamin C Heinrich, and Hans-Joachim Wuensche. Motion planning for autonomous vehicles in highly constrained urban environments. In *2016 IEEE/RSJ International Conference on Intelligent Robots and Systems (IROS)*, pages 4708–4713. IEEE, 2016.

[46] Xidong Feng, Zhepeng Cen, Jianming Hu, and Yi Zhang. Vehicle trajectory prediction using intention-based conditional variational autoencoder. In *2019 IEEE Intelligent Transportation Systems Conference (ITSC)*, pages 3514–3519. IEEE, 2019.

[47] Dániel Fényes, Balázs Németh, and Péter Gáspár. Lpv-based autonomous vehicle control using the results of big data analysis on lateral dynamics. In *2020 American Control Conference (ACC)*, pages 2250–2255. IEEE, 2020.

[48] Tharindu Fernando, Simon Denman, Sridha Sridharan, and Clinton Fookes. Deep inverse reinforcement learning for behavior prediction in autonomous driving: Accurate forecasts of vehicle motion. *IEEE Signal Processing Magazine*, 38(1):87–96, 2020.

[49] Axel Fritz and Werner Schiehlen. Nonlinear acc in simulation and measurement. *Vehicle System Dynamics*, 36(2-3):159–177, 2001.

[50] Andrei Furda and Ljubo Vlacic. Enabling safe autonomous driving in real-world city traffic using multiple criteria decision making. *IEEE Intelligent Transportation Systems Magazine*, 3(1):4–17, 2011.

[51] Bingzhao Gao, Kunyang Cai, Ting Qu, Yunfeng Hu, and Hong Chen. Personalized adaptive cruise control based on online driving style recognition technology and model predictive control. *IEEE Transactions on Vehicular Technology*, 69(11):12482–12496, 2020.

[52] Hongbo Gao, Zhen Kan, and Keqiang Li. Robust lateral trajectory following control of unmanned vehicle based on model predictive control. *IEEE/ASME Transactions on Mechatronics*, 2021.

[53] Weinan Gao, Zhong-Ping Jiang, and Kaan Ozbay. Data-driven adaptive optimal control of connected vehicles. *IEEE Transactions on Intelligent Transportation Systems*, 18(5):1122–1133, 2016.

[54] Zhen Gao, Yongchao Liang, Jiangyu Zheng, and Junyi Chen. Driving style recognition based on lane change behavior analysis using naturalistic driving data. In *CICTP 2020*, pages 4449–4461. 2020.

[55] Valentina Gatteschi, Alberto Cannavo, Fabrizio Lamberti, Lia Morra, and Paolo Montuschi. Comparing algorithms for aggressive drivingevent detection based on vehicle motion data. *IEEE Transactions on Vehicular Technology*, 2021.

[56] Mohammad Goli and Azim Eskandarian. Merging strategies, trajectory planning and controls for platoon of connected, and autonomous vehicles. *International Journal of Intelligent Transportation Systems Research*, 18(1):153–173, 2020.

[57] Franz Gritschneder, Patrick Hatzelmann, Markus Thom, Felix Kunz, and Klaus Dietmayer. Adaptive learning based on guided exploration for decision making at roundabouts. In *2016 IEEE Intelligent Vehicles Symposium (IV)*, pages 433–440. IEEE, 2016.

[58] Tianyu Gu and John M Dolan. Toward human-like motion planning in urban environments. In *2014 IEEE Intelligent Vehicles Symposium Proceedings*, pages 350–355. IEEE, 2014.

[59] Xinping Gu, Yunpeng Han, and Junfu Yu. A novel lane-changing decision model for autonomous vehicles based on deep autoencoder network and xgboost. *IEEE Access*, 8:9846–9863, 2020.

[60] Yanlei Gu, Yoriyoshi Hashimoto, Li-Ta Hsu, Miho Iryo-Asano, and Shunsuke Kamijo. Human-like motion planning model for driving in signalized intersections. *IATSS Research*, 41(3):129–139, 2017.

[61] Chen Guangfeng, Zhai Linlin, Huang Qingqing, Li Lei, and Shi Jiawen. Trajectory planning of delta robot for fixed point pick and placement. In *2012 Fourth International Symposium on Information Science and Engineering*, pages 236–239. IEEE, 2012.

[62] Carlos Guardiola, Benjamin Pla, David Blanco-Rodríguez, and A Reig. Modelling driving behaviour and its impact on the energy management problem in hybrid electric vehicles. *International Journal of Computer Mathematics*, 91(1):147–156, 2014.

[63] Mahir Gulzar, Yar Muhammad, and Naveed Muhammad. A survey on motion prediction of pedestrians and vehicles for autonomous driving. *IEEE Access*, 2021.

[64] Chunzhao Guo, Kiyosumi Kidono, Ryuta Terashima, and Yoshiko Kojima. Toward human-like behavior generation in urban environment based on markov decision process with hybrid potential maps. In *2018 IEEE Intelligent Vehicles Symposium (IV)*, pages 2209–2215. IEEE, 2018.

[65] Chunzhao Guo, Takashi Owaki, Kiyosumi Kidono, Takashi Machida, Ryuta Terashima, and Yoshiko Kojima. Toward human-like lane following behavior in urban environment with a learning-based behavior-induction potential map. In *2017 IEEE International Conference on Robotics and Automation (ICRA)*, pages 1409–1416. IEEE, 2017.

[66] Jinghua Guo, Yugong Luo, Keqiang Li, and Yifan Dai. Coordinated path-following and direct yaw-moment control of autonomous electric vehicles with sideslip angle estimation. *Mechanical Systems and Signal Processing*, 105:183–199, 2018.

[67] Qiuyi Guo, Zhiguo Zhao, Peihong Shen, Xiaowen Zhan, and Jingwei Li. Adaptive optimal control based on driving style recognition for plug-in hybrid electric vehicle. *Energy*, 186:115824, 2019.

[68] Ibrahim A Hameed. Coverage path planning software for autonomous robotic lawn mower using dubins' curve. In *2017 IEEE International Conference on Real-time Computing and Robotics (RCAR)*, pages 517–522. IEEE, 2017.

[69] Wei Han, Wenshuo Wang, Xiaohan Li, and Junqiang Xi. Statistical-based approach for driving style recognition using bayesian probability with kernel density estimation. *IET Intelligent Transport Systems*, 13(1):22–30, 2019.

[70] Yu Han, Qidan Zhu, and Yao Xiao. Data-driven control of autonomous vehicle using recurrent fuzzy neural network combined with pid method. In *2018 37th Chinese Control Conference (CCC)*, pages 5239–5244. IEEE, 2018.

[71] Peng Hang and Xinbo Chen. Integrated chassis control algorithm design for path tracking based on four-wheel steering and direct yaw-moment control. *Proceedings of the Institution of Mechanical Engineers, Part I: Journal of Systems and Control Engineering*, 233(6):625–641, 2019.

[72] Peng Hang, Xinbo Chen, Shude Fang, and Fengmei Luo. Robust control for four-wheel-independent-steering electric vehicle with steer-by-wire system. *International Journal of Automotive Technology*, 18(5):785–797, 2017.

[73] Peng Hang, Xinbo Chen, and Fengmei Luo. Lpv/hinf controller design for path tracking of autonomous ground vehicles through four-wheel steering and direct yaw-moment control. *International Journal of Automotive Technology*, 20(4):679–691, 2019.

[74] Peng Hang, Xinbo Chen, Bang Zhang, and Tingju Tang. Longitudinal velocity tracking control of a 4wid electric vehicle. *IFAC-PapersOnLine*, 51(31):790–795, 2018.

[75] Peng Hang, Chao Huang, Zhongxu Hu, Yang Xing, and Chen Lv. Decision making of connected automated vehicles at an unsignalized roundabout considering personalized driving behaviours. *IEEE Transactions on Vehicular Technology*, 70(5):4051–4064, 2021.

[76] Peng Hang, Sunan Huang, Xinbo Chen, and Kok Kiong Tan. Path planning of collision avoidance for unmanned ground vehicles: A nonlinear model predictive control approach. *Proceedings of the Institution of Mechanical Engineers, Part I: Journal of Systems and Control Engineering*, 235(2):222–236, 2021.

[77] Peng Hang, Fengmei Luo, Shude Fang, and Xinbo Chen. Path tracking control of a four-wheel-independent-steering electric vehicle based on model predictive control. In *2017 36th Chinese Control Conference (CCC)*, pages 9360–9366. IEEE, 2017.

[78] Peng Hang, Chen Lv, Chao Huang, Jiacheng Cai, Zhongxu Hu, and Yang Xing. An integrated framework of decision making and motion planning for autonomous vehicles considering social behaviors. *IEEE Transactions on Vehicular Technology*, 69(12):14458–14469, 2020.

[79] Peng Hang, Chen Lv, Yang Xing, Chao Huang, and Zhongxu Hu. Human-like decision making for autonomous driving: A noncooperative game theoretic approach. *IEEE Transactions on Intelligent Transportation Systems*, 22(4):2076–2087, 2020.

[80] Peng Hang, Xin Xia, Guang Chen, and Xinbo Chen. Active safety control of automated electric vehicles at driving limits: A tube-based mpc approach. *IEEE Transactions on Transportation Electrification*, 2021.

[81] Terry Harris. Credit scoring using the clustered support vector machine. *Expert Systems with Applications*, 42(2):741–750, 2015.

[82] Xiangkun He, Kaiming Yang, Yulong Liu, and Xuewu Ji. A novel direct yaw moment control system for autonomous vehicle. Technical report, SAE Technical Paper, 2018.

[83] Toshihiro Hiraoka, Osamu Nishihara, and Hiromitsu Kumamoto. Automatic path-tracking controller of a four-wheel steering vehicle. *Vehicle System Dynamics*, 47(10):1205–1227, 2009.

[84] L Hu, ZQ Gu, J Huang, Y Yang, and X Song. Research and realization of optimum route planning in vehicle navigation systems based on a hybrid genetic algorithm. *Proceedings of the Institution of Mechanical Engineers, Part D: Journal of Automobile Engineering*, 222(5):757–763, 2008.

[85] Xiangwang Hu and Jian Sun. Trajectory optimization of connected and autonomous vehicles at a multilane freeway merging area. *Transportation Research Part C: Emerging Technologies*, 101:111–125, 2019.

[86] Xuemin Hu, Long Chen, Bo Tang, Dongpu Cao, and Haibo He. Dynamic path planning for autonomous driving on various roads with avoidance of static and moving obstacles. *Mechanical Systems and Signal Processing*, 100:482–500, 2018.

[87] Chao Huang, Hailong Huang, Junzhi Zhang, Peng Hang, Zhongxu Hu, and Chen Lv. Human-machine cooperative trajectory planning and tracking for safe automated driving. *IEEE Transactions on Intelligent Transportation Systems*, 2021.

[88] Chao Huang, Chen Lv, Peng Hang, and Yang Xing. Toward safe and personalized autonomous driving: Decision-making and motion control with dpf and cdt techniques. *IEEE/ASME Transactions on Mechatronics*, 26(2):611–620, 2021.

[89] Tianyu Huang and Zhanbo Sun. Cooperative ramp merging for mixed traffic with connected automated vehicles and human-operated vehicles. *IFAC-PapersOnLine*, 52(24):76–81, 2019.

[90] Xin Huang, Stephen G McGill, Jonathan A DeCastro, Luke Fletcher, John J Leonard, Brian C Williams, and Guy Rosman. Diversitygan: Diversity-aware vehicle motion prediction via latent semantic sampling. *IEEE Robotics and Automation Letters*, 5(4):5089–5096, 2020.

[91] Yanjun Huang, Hong Wang, Amir Khajepour, Haitao Ding, Kang Yuan, and Yechen Qin. A novel local motion planning framework for autonomous vehicles based on resistance network and model predictive control. *IEEE Transactions on Vehicular Technology*, 69(1):55–66, 2019.

[92] Zhiyu Huang, Xiaoyu Mo, and Chen Lv. Multi-modal motion prediction with transformer-based neural network for autonomous driving. *arXiv preprint arXiv:2109.06446*, 2021.

[93] Zhiyu Huang, Jingda Wu, and Chen Lv. Driving behavior modeling using naturalistic human driving data with inverse reinforcement learning. *IEEE Transactions on Intelligent Transportation Systems*, 2021.

[94] Yuji Ito, Md Abdus Samad Kamal, Takayoshi Yoshimura, and Shun-ichi Azuma. Coordination of connected vehicles on merging roads using pseudo-perturbation-based broadcast control. *IEEE Transactions on Intelligent Transportation Systems*, 20(9):3496–3512, 2018.

[95] Arash Jahangiri, Vincent J Berardi, and Sahar Ghanipoor Machiani. Application of real field connected vehicle data for aggressive driving identification on horizontal curves. *IEEE Transactions on Intelligent Transportation Systems*, 19(7):2316–2324, 2017.

[96] Yongsu Jeon, Chanwoo Kim, Hyunwook Lee, and Yunju Baek. Real-time aggressive driving detection system based on in-vehicle information using lora communication. In *MATEC Web of Conferences*, volume 308, page 06001. EDP Sciences, 2020.

[97] Yonghwan Jeong. Self-adaptive motion prediction-based proactive motion planning for autonomous driving in urban environments. *IEEE Access*, 9:105612–105626, 2021.

[98] Yonghwan Jeong, Seonwook Kim, and Kyongsu Yi. Surround vehicle motion prediction using lstm-rnn for motion planning of autonomous vehicles at multi-lane turn intersections. *IEEE Open Journal of Intelligent Transportation Systems*, 1:2–14, 2020.

[99] Yonghwan Jeong and Kyongsu Yi. Target vehicle motion prediction-based motion planning framework for autonomous driving in uncontrolled intersections. *IEEE Transactions on Intelligent Transportation Systems*, 2019.

[100] Shuo Jia, Fei Hui, Shining Li, Xiangmo Zhao, and Asad J Khattak. Long short-term memory and convolutional neural network for abnormal driving behaviour recognition. *IET Intelligent Transport Systems*, 14(5):306–312, 2020.

[101] Chaoyang Jiang, Hanqing Tian, Jibin Hu, Jiankun Zhai, Chao Wei, and Jun Ni. Learning based predictive error estimation and compensator design for autonomous vehicle path tracking. In *2020 15th IEEE Conference on Industrial Electronics and Applications (ICIEA)*, pages 1496–1500. IEEE, 2020.

[102] Le Jiang, Yafei Wang, Lin Wang, and Jingkai Wu. Path tracking control based on deep reinforcement learning in autonomous driving. In *2019 3rd Conference on Vehicle Control and Intelligence (CVCI)*, pages 1–6. IEEE, 2019.

[103] Lu Jianglin. Intelligent vehicle automatic lane changing lateral control method based on deep learning. In *2020 IEEE International Conference on Industrial Application of Artificial Intelligence (IAAI)*, pages 278–283. IEEE, 2020.

[104] Shoucai Jing, Fei Hui, Xiangmo Zhao, Jackeline Rios-Torres, and Asad J Khattak. Cooperative game approach to optimal merging sequence and on-ramp merging control of connected and automated vehicles. *IEEE Transactions on Intelligent Transportation Systems*, 20(11):4234–4244, 2019.

[105] Zhang Jinzhu and Zhang Hongtian. Vehicle stability control based on adaptive pid control with single neuron network. In *2010 2nd International Asia Conference on Informatics in Control, Automation and Robotics (CAR 2010)*, volume 1, pages 434–437. IEEE, 2010.

[106] Derick A Johnson and Mohan M Trivedi. Driving style recognition using a smartphone as a sensor platform. In *2011 14th International IEEE Conference on Intelligent Transportation Systems (ITSC)*, pages 1609–1615. IEEE, 2011.

[107] Wang Jun, Qingnian Wang, Peng-yu Wang, et al. Adaptive shift control strategy based on driving style recognition. Technical report, SAE Technical Paper, 2013.

[108] Nidhi Kalra, Raman Kumar Goyal, Anshu Parashar, Jaskirat Singh, and Gagan Singla. Driving style recognition system using smartphone sensors based on fuzzy logic. *CMC-COMPUTERS MATERIALS & CONTINUA*, 69(2):1967–1978, 2021.

[109] Md Abdus Samad Kamal, Shun Taguchi, and Takayoshi Yoshimura. Efficient driving on multilane roads under a connected vehicle environment. *IEEE Transactions on Intelligent Transportation Systems*, 17(9):2541–2551, 2016.

[110] Kyungwon Kang and Hesham A Rakha. Modeling driver merging behavior: a repeated game theoretical approach. *Transportation research record*, 2672(20):144–153, 2018.

[111] Liuwang Kang and Haiying Shen. A reinforcement learning based decision-making system with aggressive driving behavior consideration for autonomous vehicles. In *2021 18th Annual IEEE International Conference on Sensing, Communication, and Networking (SECON)*, pages 1–9. IEEE, 2021.

[112] Nadezda Karginova, Stefan Byttner, and Magnus Svensson. Data-driven methods for classification of driving styles in buses. *SAE Tech. Pap*, 2012.

[113] Yasser H Khalil and Hussein T Mouftah. Licanext: Incorporating sequential range residuals for additional advancement in joint perception and motion prediction. *IEEE Access*, 2021.

[114] Jaehwan Kim and Dongsuk Kum. Collision risk assessment algorithm via lane-based probabilistic motion prediction of surrounding vehicles. *IEEE Transactions on Intelligent Transportation Systems*, 19(9):2965–2976, 2017.

[115] Taewan Kim and H Jin Kim. Path tracking control and identification of tire parameters using on-line model-based reinforcement learning. In *2016 16th International Conference on Control, Automation and Systems (ICCAS)*, pages 215–219. IEEE, 2016.

[116] Jordanka Kovaceva, Irene Isaksson-Hellman, and Nikolce Murgovski. Identification of aggressive driving from naturalistic data in car-following situations. *Journal of safety research*, 73:225–234, 2020.

[117] Omurcan Kumtepe, Gozde Bozdagi Akar, and Enes Yuncu. Driver aggressiveness detection using visual information from forward camera. In *2015 12th IEEE International Conference on Advanced Video and Signal Based Surveillance (AVSS)*, pages 1–6. IEEE, 2015.

[118] Omurcan Kumtepe, Gozde Bozdagi Akar, and Enes Yuncu. Driver aggressiveness detection via multisensory data fusion. *EURASIP Journal on Image and Video Processing*, 2016(1):1–16, 2016.

[119] Shupeng Lai, Kangli Wang, Hailong Qin, Jin Q Cui, and Ben M Chen. A robust online path planning approach in cluttered environments for micro rotorcraft drones. *Control Theory and Technology*, 14(1):83–96, 2016.

[120] I Lashkov and A Kashevnik. Aggressive behavior detection based on driver heart rate and hand movement data. In *2021 IEEE International Intelligent Transportation Systems Conference (ITSC)*, pages 1490–1495. IEEE, 2021.

[121] Jooyoung Lee and Kitae Jang. A framework for evaluating aggressive driving behaviors based on in-vehicle driving records. *Transportation Research Part F: Traffic Psychology and Behaviour*, 65:610–619, 2019.

[122] Vasileios Lefkopoulos, Marcel Menner, Alexander Domahidi, and Melanie N Zeilinger. Interaction-aware motion prediction for autonomous driving: A multiple model kalman filtering scheme. *IEEE Robotics and Automation Letters*, 6(1):80–87, 2020.

[123] Yu-long Lei, Guanzheng Wen, Yao Fu, Xingzhong Li, Boning Hou, and Xiaohu Geng. Trajectory-following of a 4wid-4wis vehicle via feedforward–backstepping sliding-mode control. *Proceedings of the Institution of Mechanical Engineers, Part D: Journal of Automobile Engineering*, page 09544070211021227, 2021.

[124] Clark Letter and Lily Elefteriadou. Efficient control of fully automated connected vehicles at freeway merge segments. *Transportation Research Part C: Emerging Technologies*, 80:190–205, 2017.

[125] Aoxue Li, Haobin Jiang, Zhaojian Li, Jie Zhou, and Xinchen Zhou. Human-like trajectory planning on curved road: Learning from human drivers. *IEEE Transactions on Intelligent Transportation Systems*, 21(8):3388–3397, 2019.

[126] Aoxue Li, Haobin Jiang, Jie Zhou, and Xinchen Zhou. Learning human-like trajectory planning on urban two-lane curved roads from experienced drivers. *IEEE Access*, 7:65828–65838, 2019.

[127] Boyuan Li, Haiping Du, and Weihua Li. Trajectory control for autonomous electric vehicles with in-wheel motors based on a dynamics model approach. *IET Intelligent Transport Systems*, 10(5):318–330, 2016.

[128] Dachuan Li, Yunjiang Wu, Bing Bai, and Qi Hao. Behavior and interaction-aware motion planning for autonomous driving vehicles based on hierarchical intention and motion prediction. In *2020 IEEE 23rd International Conference on Intelligent Transportation Systems (ITSC)*, pages 1–8. IEEE, 2020.

[129] Jianqiang Li, Genqiang Deng, Chengwen Luo, Qiuzhen Lin, Qiao Yan, and Zhong Ming. A hybrid path planning method in unmanned air/ground vehicle (uav/ugv) cooperative systems. *IEEE Transactions on Vehicular Technology*, 65(12):9585–9596, 2016.

[130] Junjun Li, Bowei Xu, Yongsheng Yang, and Huafeng Wu. Three-phase qubits-based quantum ant colony optimization algorithm for path planning of automated guided vehicles. *Int. J. Robot. Autom*, 34(2):156–163, 2019.

[131] Liangzhi Li, Kaoru Ota, and Mianxiong Dong. Humanlike driving: Empirical decision-making system for autonomous vehicles. *IEEE Transactions on Vehicular Technology*, 67(8):6814–6823, 2018.

[132] Mingxing Li and Yingmin Jia. Velocity tracking control based on throttle-pedal-moving data mapping for the autonomous vehicle. In *Chinese Intelligent Systems Conference*, pages 690–698. Springer, 2019.

[133] Shengbo Eben Li, Feng Gao, Keqiang Li, Le-Yi Wang, Keyou You, and Dongpu Cao. Robust longitudinal control of multi-vehicle systemsa distributed h-infinity method. *IEEE Transactions on Intelligent Transportation Systems*, 19(9):2779–2788, 2017.

[134] Yiyang Li, Chiyomi Miyajima, Norihide Kitaoka, and Kazuya Takeda. Evaluation method for aggressiveness of driving behavior using drive recorders. *IEEJ Journal of Industry Applications*, 4(1):59–66, 2015.

[135] Zhihui Li, Cong Wu, Pengfei Tao, Jing Tian, and Lin Ma. Dp and ds-lcd: A new lane change decision model coupling drivers psychology and driving style. *IEEE Access*, 8:132614–132624, 2020.

[136] Fen Lin, Yaowen Zhang, Youqun Zhao, Guodong Yin, Huiqi Zhang, and Kaizheng Wang. Trajectory tracking of autonomous vehicle with the fusion of dyc and longitudinal–lateral control. *Chinese Journal of Mechanical Engineering*, 32(1):1–16, 2019.

[137] Qingfeng Lin, Shengbo Eben Li, Xuejin Du, Xiaowu Zhang, Huei Peng, Yugong Luo, and Keqiang Li. Minimize the fuel consumption of connected vehicles between two red-signalized intersections in urban traffic. *IEEE Transactions on Vehicular Technology*, 67(10):9060–9072, 2018.

[138] Manuel Lindorfer, Christoph F Mecklenbraeuker, and Gerald Ostermayer. Modeling the imperfect driver: Incorporating human factors in a microscopic traffic model. *IEEE Transactions on Intelligent Transportation Systems*, 19(9), 2018.

[139] Changliu Liu, Chung-Wei Lin, Shinichi Shiraishi, and Masayoshi Tomizuka. Distributed conflict resolution for connected autonomous vehicles. *IEEE Transactions on Intelligent Vehicles*, 3(1):18–29, 2017.

[140] Qinghe Liu, Lijun Zhao, Zhibin Tan, and Wen Chen. Global path planning for autonomous vehicles in off-road environment via an a-star algorithm. *International Journal of Vehicle Autonomous Systems*, 13(4):330–339, 2017.

[141] Runqiao Liu, Minxiang Wei, Nan Sang, and Jianwei Wei. Research on curved path tracking control for four-wheel steering vehicle considering road adhesion coefficient. *Mathematical Problems in Engineering*, 2020, 2020.

[142] Runqiao Liu, Minxiang Wei, and Wanzhong Zhao. Trajectory tracking control of four wheel steering under high speed emergency obstacle avoidance. *International Journal of Vehicle Design*, 77(1-2):1–21, 2018.

[143] Wei Liu, Zhiheng Li, Li Li, and Fei-Yue Wang. Parking like a human: A direct trajectory planning solution. *IEEE Transactions on Intelligent Transportation Systems*, 18(12):3388–3397, 2017.

[144] Yonggang Liu, Jiming Wang, Pan Zhao, Datong Qin, and Zheng Chen. Research on classification and recognition of driving styles based on feature engineering. *IEEE Access*, 7:89245–89255, 2019.

[145] Yonggang Liu, Xiao Wang, Liang Li, Shuo Cheng, and Zheng Chen. A novel lane change decision-making model of autonomous vehicle based on support vector machine. *IEEE Access*, 7:26543–26550, 2019.

[146] Zijun Liu, Shuo Cheng, Xuewu Ji, Liang Li, and Lingtao Wei. A hierarchical anti-disturbance path tracking control scheme for autonomous vehicles under complex driving conditions. *IEEE Transactions on Vehicular Technology*, 2021.

[147] Chen Lv, Xiaosong Hu, Alberto Sangiovanni-Vincentelli, Yutong Li, Clara Marina Martinez, and Dongpu Cao. Driving-style-based codesign optimization of an automated electric vehicle: A cyber-physical system approach. *IEEE Transactions on Industrial Electronics*, 66(4):2965–2975, 2018.

[148] Jun Ma, Zilong Cheng, Xiaoxue Zhang, Abdullah Al Mamun, Clarence W de Silva, and Tong Heng Lee. Data-driven predictive control for multi-agent decision making with chance constraints. *arXiv preprint arXiv:2011.03213*, 2020.

[149] Jun Ma, Zilong Cheng, Xiaoxue Zhang, Masayoshi Tomizuka, and Tong Heng Lee. Optimal decentralized control for uncertain systems by symmetric gauss-seidel semi-proximal alm. *IEEE Transactions on Automatic Control*, 66(11):5554–5560, 2021.

[150] Yong Ma, Mengqi Hu, and Xinping Yan. Multi-objective path planning for unmanned surface vehicle with currents effects. *ISA Transactions*, 75:137–156, 2018.

[151] Yongfeng Ma, Kun Tang, Shuyan Chen, Aemal J Khattak, and Yingjiu Pan. On-line aggressive driving identification based on in-vehicle kinematic parameters under naturalistic driving conditions. *Transportation Research Part C: Emerging Technologies*, 114:554–571, 2020.

[152] Yongfeng Ma, Ziyu Zhang, Shuyan Chen, Yanan Yu, and Kun Tang. A comparative study of aggressive driving behavior recognition algorithms based on vehicle motion data. *IEEE Access*, 7:8028–8038, 2018.

[153] Vivek Mahalingam and Abhishek Agrawal. Learning agents based intelligent transport and routing systems for autonomous vehicles and their respective vehicle control systems based on model predictive control (mpc). In *2016 IEEE International Conference on Recent Trends in Electronics, Information & Communication Technology (RTEICT)*, pages 284–290. IEEE, 2016.

[154] Sampurna Mandal, Swagatam Biswas, Valentina E Balas, Rabindra Nath Shaw, and Ankush Ghosh. Motion prediction for autonomous vehicles from lyft dataset using deep learning. In *2020 IEEE 5th international conference on computing communication and automation (ICCCA)*, pages 768–773. IEEE, 2020.

[155] Charles Marks, Arash Jahangiri, and Sahar Ghanipoor Machiani. Iterative dbscan (i-dbscan) to identify aggressive driving behaviors within unlabeled real-world driving data. In *2019 IEEE Intelligent Transportation Systems Conference (ITSC)*, pages 2324–2329. IEEE, 2019.

[156] Fabio Martinelli, Francesco Mercaldo, Albina Orlando, Vittoria Nardone, Antonella Santone, and Arun Kumar Sangaiah. Human behavior characterization for driving style recognition in vehicle system. *Computers & Electrical Engineering*, 83:102504, 2020.

[157] Clara Marina Martinez, Mira Heucke, Fei-Yue Wang, Bo Gao, and Dongpu Cao. Driving style recognition for intelligent vehicle control and advanced driver assistance: A survey. *IEEE Transactions on Intelligent Transportation Systems*, 19(3):666–676, 2017.

[158] Behrooz Mashadi and Majid Majidi. Two-phase optimal path planning of autonomous ground vehicles using pseudo-spectral method. *Proceedings of the Institution of Mechanical Engineers, Part K: Journal of Multi-body Dynamics*, 228(4):426–437, 2014.

[159] Joel C McCall and Mohan M Trivedi. Driver behavior and situation aware brake assistance for intelligent vehicles. *Proceedings of the IEEE*, 95(2):374–387, 2007.

[160] Haigen Min, Yiming Yang, Yukun Fang, Pengpeng Sun, and Xiangmo Zhao. Constrained optimization and distributed model predictive control-based merging strategies for adjacent connected autonomous vehicle platoons. *IEEE Access*, 7:163085–163096, 2019.

[161] Xiaoyu Mo, Yang Xing, and Chen Lv. Interaction-aware trajectory prediction of connected vehicles using cnn-lstm networks. In *IECON 2020 The 46th Annual Conference of the IEEE Industrial Electronics Society*, pages 5057–5062. IEEE, 2020.

[162] Xiaoyu Mo, Yang Xing, and Chen Lv. Recog: A deep learning framework with heterogeneous graph for interaction-aware trajectory prediction. *arXiv preprint arXiv:2012.05032*, 2020.

[163] Xiaoyu Mo, Yang Xing, and Chen Lv. Heterogeneous edge-enhanced graph attention network for multi-agent trajectory prediction. *arXiv preprint arXiv:2106.07161*, 2021.

[164] Sangwoo Moon, Eunmi Oh, and David Hyunchul Shim. An integral framework of task assignment and path planning for multiple unmanned aerial vehicles in dynamic environments. *Journal of Intelligent & Robotic Systems*, 70(1):303–313, 2013.

[165] Seungwuk Moon and Kyongsu Yi. Human driving data-based design of a vehicle adaptive cruise control algorithm. *Vehicle System Dynamics*, 46(8):661–690, 2008.

[166] Youness Moukafih, Hakim Hafidi, and Mounir Ghogho. Aggressive driving detection using deep learning-based time series classification. In *2019 IEEE International Symposium on INnovations in Intelligent SysTems and Applications (INISTA)*, pages 1–5. IEEE, 2019.

[167] Freddy Antony Mullakkal-Babu, Meng Wang, Bart van Arem, Barys Shyrokau, and Riender Happee. A hybrid submicroscopic-microscopic traffic flow simulation framework. *IEEE Transactions on Intelligent Transportation Systems*, 2020.

[168] Xiaoxiang Na and David J Cole. Game-theoretic modeling of the steering interaction between a human driver and a vehicle collision avoidance controller. *IEEE Transactions on Human-Machine Systems*, 45(1):25–38, 2014.

[169] Xiaoxiang Na and David J Cole. Application of open-loop stackelberg equilibrium to modeling a driver's interaction with vehicle active steering control in obstacle avoidance. *IEEE Transactions on Human-Machine Systems*, 47(5):673–685, 2017.

[170] Daniel Chi Kit Ngai and Nelson Hon Ching Yung. A multiple-goal reinforcement learning method for complex vehicle overtaking maneuvers. *IEEE Transactions on Intelligent Transportation Systems*, 12(2):509–522, 2011.

[171] Anh-Tu Nguyen, Thierry-Marie Guerra, Jagat Rath, Hui Zhang, and Reinaldo Palhares. Set-invariance based fuzzy output tracking control for vehicle autonomous driving under uncertain lateral forces and steering constraints. In *2020 IEEE International Conference on Fuzzy Systems (FUZZ-IEEE)*, pages 1–7. IEEE, 2020.

[172] Jianqiang Nie, Jian Zhang, Wanting Ding, Xia Wan, Xiaoxuan Chen, and Bin Ran. Decentralized cooperative lane-changing decision-making for connected autonomous vehicles. *IEEE Access*, 4:9413–9420, 2016.

[173] Vladimir Nikulin. Driving style identification with unsupervised learning. In *International Conference on Machine Learning and Data Mining in Pattern Recognition*, pages 155–169. Springer, 2016.

[174] Julia Nilsson, Mattias Brännström, Jonas Fredriksson, and Erik Coelingh. Longitudinal and lateral control for automated yielding maneuvers. *IEEE Transactions on Intelligent Transportation Systems*, 17(5):1404–1414, 2016.

[175] Samyeul Noh. Decision-making framework for autonomous driving at road intersections: Safeguarding against collision, overly conservative behavior, and violation vehicles. *IEEE Transactions on Industrial Electronics*, 66(4):3275–3286, 2018.

[176] Bunyo Okumura, Michael R James, Yusuke Kanzawa, Matthew Derry, Katsuhiro Sakai, Tomoki Nishi, and Danil Prokhorov. Challenges in perception and decision making for intelligent automotive vehicles: A case study. *IEEE Transactions on Intelligent Vehicles*, 1(1):20–32, 2016.

[177] Atakan Ondoğan and Hasan Serhan Yavuz. Fuzzy logic based adaptive cruise control for low-speed following. In *2019 3rd International Symposium on Multidisciplinary Studies and Innovative Technologies (ISMSIT)*, pages 1–5. IEEE, 2019.

[178] Parham Pahlavani and Mahmoud R Delavar. Multi-criteria route planning based on a drivers preferences in multi-criteria route selection. *Transportation Research Part C: Emerging Technologies*, 40:14–35, 2014.

[179] Shuang Pan, Yafei Wang, and Kaizheng Wang. A game theory-based model predictive controller for mandatory lane change of multiple vehicles. In *2020 4th CAA International Conference on Vehicular Control and Intelligence (CVCI)*, pages 731–736. IEEE, 2020.

[180] Prachi Pardhi, Kiran Yadav, Siddhansh Shrivastav, Satya Prakash Sahu, and Deepak Kumar Dewangan. Vehicle motion prediction for autonomous navigation system using 3 dimensional convolutional neural network. In *2021 5th*

International Conference on Computing Methodologies and Communication (ICCMC), pages 1322–1329. IEEE, 2021.

[181] Sasinee Pruekprasert, Jérémy Dubut, Xiaoyi Zhang, Chao Huang, and Masako Kishida. A game-theoretic approach to decision making for multiple vehicles at roundabout. *arXiv preprint arXiv:1904.06224*, 2019.

[182] Geqi Qi, Jianping Wu, Yang Zhou, Yiman Du, Yuhan Jia, Nick Hounsell, and Neville A Stanton. Recognizing driving styles based on topic models. *Transportation Research Part D: Transport and Environment*, 66:13–22, 2019.

[183] Yalda Rahmati and Alireza Talebpour. Towards a collaborative connected, automated driving environment: A game theory based decision framework for unprotected left turn maneuvers. In *2017 IEEE Intelligent Vehicles Symposium (IV)*, pages 1316–1321. IEEE, 2017.

[184] Yadollah Rasekhipour, Amir Khajepour, Shih-Ken Chen, and Bakhtiar Litkouhi. A potential field-based model predictive path-planning controller for autonomous road vehicles. *IEEE Transactions on Intelligent Transportation Systems*, 18(5):1255–1267, 2016.

[185] Jackeline Rios-Torres and Andreas A Malikopoulos. Automated and cooperative vehicle merging at highway on-ramps. *IEEE Transactions on Intelligent Transportation Systems*, 18(4):780–789, 2016.

[186] Nicholas Rizzo, Ethan Sprissler, Yuan Hong, and Sanjay Goel. Privacy preserving driving style recognition. In *2015 International Conference on Connected Vehicles and Expo (ICCVE)*, pages 232–237. IEEE, 2015.

[187] Maradona Rodrigues, Andrew McGordon, Graham Gest, and James Marco. Autonomous navigation in interaction-based environmentsa case of non-signalized roundabouts. *IEEE Transactions on Intelligent Vehicles*, 3(4):425–438, 2018.

[188] Mohammad Rokonuzzaman, Navid Mohajer, Saeid Nahavandi, and Shady Mohamed. Learning-based model predictive control for path tracking control of autonomous vehicle. In *2020 IEEE International Conference on Systems, Man, and Cybernetics (SMC)*, pages 2913–2918. IEEE, 2020.

[189] Thomas Rosenstatter and Cristofer Englund. Modelling the level of trust in a cooperative automated vehicle control system. *IEEE Transactions on Intelligent Transportation Systems*, 19(4):1237–1247, 2017.

[190] Fridulv Sagberg, Selpi, Giulio Francesco Bianchi Piccinini, and Johan Engström. A review of research on driving styles and road safety. *Human factors*, 57(7):1248–1275, 2015.

[191] Basak Sakcak, Luca Bascetta, and Gianni Ferretti. Human-like path planning in the presence of landmarks. In *International Workshop on Modelling and Simulation for Autonomous Systems*, pages 281–287. Springer, 2016.

[192] Subramanian Saravanakumar and Thondiyath Asokan. Multipoint potential field method for path planning of autonomous underwater vehicles in 3d space. *Intelligent Service Robotics*, 6(4):211–224, 2013.

[193] Keisuke Sato. Deadlock-free motion planning using the laplace potential field. *Advanced Robotics*, 7(5):449–461, 1992.

[194] Christoph Schöller, Vincent Aravantinos, Florian Lay, and Alois Knoll. What the constant velocity model can teach us about pedestrian motion prediction. *IEEE Robotics and Automation Letters*, 5(2):1696–1703, 2020.

[195] Robin Schubert. Evaluating the utility of driving: Toward automated decision making under uncertainty. *IEEE Transactions on Intelligent Transportation Systems*, 13(1):354–364, 2011.

[196] Alexander Schwab and Jan Lunze. Vehicle platooning and cooperative merging. *IFAC-PapersOnLine*, 52(5):353–358, 2019.

[197] Bin Shi, Li Xu, Jie Hu, Yun Tang, Hong Jiang, Wuqiang Meng, and Hui Liu. Evaluating driving styles by normalizing driving behavior based on personalized driver modeling. *IEEE Transactions on Systems, Man, and Cybernetics: Systems*, 45(12):1502–1508, 2015.

[198] T Shiotsuka, A Nagamatsu, and K Yoshida. Adaptive control of 4ws system by using neural network. *Vehicle System Dynamics*, 22(5-6):411–424, 1993.

[199] Hamed Shorakaei, Mojtaba Vahdani, Babak Imani, and Ali Gholami. Optimal cooperative path planning of unmanned aerial vehicles by a parallel genetic algorithm. *Robotica*, 34(4):823–836, 2016.

[200] Ankur Sinha, Pekka Malo, and Kalyanmoy Deb. Efficient evolutionary algorithm for single-objective bilevel optimization. *arXiv preprint arXiv:1303.3901*, 2013.

[201] Jeffrey L Solka, James C Perry, Brian R Poellinger, and George W Rogers. Fast computation of optimal paths using a parallel dijkstra algorithm with embedded constraints. *Neurocomputing*, 8(2):195–212, 1995.

[202] Amir R Soltani, Hissam Tawfik, John Yannis Goulermas, and Terrence Fernando. Path planning in construction sites: performance evaluation of the dijkstra, a*, and ga search algorithms. *Advanced Engineering Informatics*, 16(4):291–303, 2002.

[203] Bohua Sun, Weiwen Deng, Jian Wu, Yaxin Li, Bing Zhu, and Liguang Wu. Research on the classification and identification of drivers driving style. In *2017 10th International Symposium on Computational Intelligence and Design (ISCID)*, volume 1, pages 28–32. IEEE, 2017.

[204] Chuanyang Sun, Xin Zhang, Quan Zhou, and Ying Tian. A model predictive controller with switched tracking error for autonomous vehicle path tracking. *IEEE Access*, 7:53103–53114, 2019.

[205] Yiwen Sun, Kun Fu, Zheng Wang, Donghua Zhou, Kailun Wu, Jieping Ye, and Changshui Zhang. Codriver eta: Combine driver information in estimated time of arrival by driving style learning auxiliary task. *IEEE Transactions on Intelligent Transportation Systems*, 2020.

[206] Evženie Suzdaleva and Ivan Nagy. Two-layer pointer model of driving style depending on the driving environment. *Transportation Research Part B: methodological*, 128:254–270, 2019.

[207] Mikhail Sysoev, Andrej Kos, Jože Guna, and Matevž Pogačnik. Estimation of the driving style based on the users activity and environment influence. *Sensors*, 17(10):2404, 2017.

[208] Mehrdad Tajalli and Ali Hajbabaie. Distributed optimization and coordination algorithms for dynamic speed optimization of connected and autonomous vehicles in urban street networks. *Transportation Research Part C: Emerging Technologies*, 95:497–515, 2018.

[209] Alireza Talebpour, Hani S Mahmassani, and Samer H Hamdar. Modeling lane-changing behavior in a connected environment: A game theory approach. *Transportation Research Procedia*, 7:420–440, 2015.

[210] Luqi Tang, Fuwu Yan, Bin Zou, Kewei Wang, and Chen Lv. An improved kinematic model predictive control for high-speed path tracking of autonomous vehicles. *IEEE Access*, 8:51400–51413, 2020.

[211] Orit Taubman-Ben-Ari and Dalia Yehiel. Driving styles and their associations with personality and motivation. *Accident Analysis & Prevention*, 45:416–422, 2012.

[212] Ran Tian, Nan Li, Ilya Kolmanovsky, Yildiray Yildiz, and Anouck R Girard. Game-theoretic modeling of traffic in unsignalized intersection network for autonomous vehicle control verification and validation. *IEEE Transactions on Intelligent Transportation Systems*, 2020.

[213] Ran Tian, Sisi Li, Nan Li, Ilya Kolmanovsky, Anouck Girard, and Yildiray Yildiz. Adaptive game-theoretic decision making for autonomous vehicle control at roundabouts. In *2018 IEEE Conference on Decision and Control (CDC)*, pages 321–326. IEEE, 2018.

[214] Xiang Tian, Yingfeng Cai, Xiaodong Sun, Zhen Zhu, and Yiqiang Xu. An adaptive ecms with driving style recognition for energy optimization of parallel hybrid electric buses. *Energy*, 189:116151, 2019.

[215] Qiang Tu, Hui Chen, and Jiancong Li. A potential field based lateral planning method for autonomous vehicles. *SAE International Journal of Passenger Cars-Electronic and Electrical Systems*, 10(2016-01-1874):24–34, 2016.

[216] Xuyong Tu, Jingyao Gai, and Lie Tang. Robust navigation control of a 4wd/4ws agricultural robotic vehicle. *Computers and Electronics in Agriculture*, 164:104892, 2019.

[217] Vygandas Vaitkus, Paulius Lengvenis, and Gediminas Žylius. Driving style classification using long-term accelerometer information. In *2014 19th International Conference on Methods and Models in Automation and Robotics (MMAR)*, pages 641–644. IEEE, 2014.

[218] Minh Van Ly, Sujitha Martin, and Mohan M Trivedi. Driver classification and driving style recognition using inertial sensors. In *2013 IEEE Intelligent Vehicles Symposium (IV)*, pages 1040–1045. IEEE, 2013.

[219] Sandor M Veres, Levente Molnar, Nick K Lincoln, and Colin P Morice. Autonomous vehicle control systemsa review of decision making. *Proceedings of the Institution of Mechanical Engineers, Part I: Journal of Systems and Control Engineering*, 225(2):155–195, 2011.

[220] Hengyang Wang, Biao Liu, Xianyao Ping, and Quan An. Path tracking control for autonomous vehicles based on an improved mpc. *IEEE Access*, 7:161064–161073, 2019.

[221] Hong Wang, Yanjun Huang, Amir Khajepour, Yubiao Zhang, Yadollah Rasekhipour, and Dongpu Cao. Crash mitigation in motion planning for autonomous vehicles. *IEEE Transactions on Intelligent Transportation Systems*, 20(9):3313–3323, 2019.

[222] Hong Wang, Amir Khajepour, Dongpu Cao, and Teng Liu. Ethical decision making in autonomous vehicles: Challenges and research progress. *IEEE Intelligent Transportation Systems Magazine*, 2020.

[223] Hongbo Wang, Shihan Xu, and Longze Deng. Automatic lane-changing decision based on single-step dynamic game with incomplete information and collision-free path planning. In *Actuators*, volume 10, page 173. Multidisciplinary Digital Publishing Institute, 2021.

[224] Jinxiang Wang, Junmin Wang, Rongrong Wang, and Chuan Hu. A framework of vehicle trajectory replanning in lane exchanging with considerations of driver characteristics. *IEEE Transactions on Vehicular Technology*, 66(5):3583–3596, 2016.

[225] Jinxiang Wang, Guoguang Zhang, Rongrong Wang, Scott C Schnelle, and Junmin Wang. A gain-scheduling driver assistance trajectory-following algorithm considering different driver steering characteristics. *IEEE Transactions on Intelligent Transportation Systems*, 18(5):1097–1108, 2016.

[226] Meng Wang, Serge P Hoogendoorn, Winnie Daamen, Bart van Arem, and Riender Happee. Game theoretic approach for predictive lane-changing and car-following control. *Transportation Research Part C: Emerging Technologies*, 58:73–92, 2015.

[227] Rui Wang and Srdjan M Lukic. Review of driving conditions prediction and driving style recognition based control algorithms for hybrid electric vehicles. In *2011 IEEE Vehicle Power and Propulsion Conference*, pages 1–7. IEEE, 2011.

[228] Wenshuo Wang and Junqiang Xi. A rapid pattern-recognition method for driving styles using clustering-based support vector machines. In *2016 American Control Conference (ACC)*, pages 5270–5275. IEEE, 2016.

[229] Yijing Wang, Zhengxuan Liu, Zhiqiang Zuo, Zheng Li, Li Wang, and Xiaoyuan Luo. Trajectory planning and safety assessment of autonomous vehicles based on motion prediction and model predictive control. *IEEE Transactions on Vehicular Technology*, 68(9):8546–8556, 2019.

[230] YingQiao Wang. Ltn: Long-term network for long-term motion prediction. In *2021 IEEE International Intelligent Transportation Systems Conference (ITSC)*, pages 1845–1852. IEEE, 2021.

[231] Yunpeng Wang, Pinlong Cai, and Guangquan Lu. Cooperative autonomous traffic organization method for connected automated vehicles in multi-intersection road networks. *Transportation Research Part C: Emerging Technologies*, 111:458–476, 2020.

[232] Yunpeng Wang, E Wenjuan, Wenzhong Tang, Daxin Tian, Guangquan Lu, and Guizhen Yu. Automated on-ramp merging control algorithm based on internet-connected vehicles. *IET Intelligent Transport Systems*, 7(4):371–379, 2013.

[233] Zejiang Wang, Yunhao Bai, Junmin Wang, and Xiaorui Wang. Vehicle path-tracking linear-time-varying model predictive control controller parameter selection considering central process unit computational load. *Journal of Dynamic Systems, Measurement, and Control*, 141(5):051004, 2019.

[234] Stephen Waydo and Richard M Murray. Vehicle motion planning using stream functions. In *2003 IEEE International Conference on Robotics and Automation (Cat. No. 03CH37422)*, volume 2, pages 2484–2491. IEEE, 2003.

[235] Chongfeng Wei, Richard Romano, Natasha Merat, Yafei Wang, Chuan Hu, Hamid Taghavifar, Foroogh Hajiseyedjavadi, and Erwin R Boer. Risk-based autonomous vehicle motion control with considering human drivers behaviour. *Transportation Research Part C: Emerging Technologies*, 107:1–14, 2019.

[236] ChunYan Wnag, WanZhong Zhao, ZhiJiang Xu, and Guan Zhou. Path planning and stability control of collision avoidance system based on active front steering. *Science China Technological Sciences*, 60(8):1231–1243, 2017.

[237] Jian Wu, Shuo Cheng, Binhao Liu, and Congzhi Liu. A human-machine-cooperative-driving controller based on afs and dyc for vehicle dynamic stability. *Energies*, 10(11):1737, 2017.

[238] Jingda Wu, Zhiyu Huang, Peng Hang, Chao Huang, Niels De Boer, and Chen Lv. Digital twin-enabled reinforcement learning for end-to-end autonomous driving. In *2021 IEEE 1st International Conference on Digital Twins and Parallel Intelligence (DTPI)*, pages 62–65. IEEE, 2021.

[239] Pengxiang Wu, Siheng Chen, and Dimitris N Metaxas. Motionnet: Joint perception and motion prediction for autonomous driving based on bird's eye view maps. In *Proceedings of the IEEE/CVF Conference on Computer Vision and Pattern Recognition*, pages 11385–11395, 2020.

[240] Xiangfei Wu, Xin Xu, Xiaohui Li, Kai Li, and Bohan Jiang. A kernel-based extreme learning modeling method for speed decision making of autonomous land vehicles. In *2017 6th Data Driven Control and Learning Systems (DDCLS)*, pages 769–775. IEEE, 2017.

[241] Yu Wun Chai, Yoshihiro Abe, Yoshio Kano, and Masato Abe. A study on adaptation of sbw parameters to individual drivers steer characteristics for improved driver–vehicle system performance. *Vehicle System Dynamics*, 44(sup1):874–882, 2006.

[242] Wei Xiao, Lijun Zhang, and Dejian Meng. Vehicle trajectory prediction based on motion model and maneuver model fusion with interactive multiple models. *SAE International Journal of Advances and Current Practices in Mobility*, 2(2020-01-0112):3060–3071, 2020.

[243] Cheng Xiao-dong, Zhou De-yun, and Zhang Ruo-nan. New method for uav online path planning. In *2013 IEEE International Conference on Signal Processing, Communication and Computing (ICSPCC 2013)*, pages 1–5. IEEE, 2013.

[244] Yang Xing, Chen Lv, and Dongpu Cao. Personalized vehicle trajectory prediction based on joint time-series modeling for connected vehicles. *IEEE Transactions on Vehicular Technology*, 69(2):1341–1352, 2019.

[245] Yang Xing, Chen Lv, Dongpu Cao, and Chao Lu. Energy oriented driving behavior analysis and personalized prediction of vehicle states with joint time series modeling. *Applied Energy*, 261:114471, 2020.

[246] Zhichao Xing, Xingliang Liu, Rui Fang, Hui Zhang, and Zhiguang Liu. Research on qualitative classification method of drivers' driving style. In *2021 IEEE International Conference on Advances in Electrical Engineering and Computer Applications (AEECA)*, pages 709–716. IEEE, 2021.

[247] Chengke Xiong, Danfeng Chen, Di Lu, Zheng Zeng, and Lian Lian. Path planning of multiple autonomous marine vehicles for adaptive sampling using voronoi-based ant colony optimization. *Robotics and Autonomous Systems*, 115:90–103, 2019.

[248] Biao Xu, Shengbo Eben Li, Yougang Bian, Shen Li, Xuegang Jeff Ban, Jianqiang Wang, and Keqiang Li. Distributed conflict-free cooperation for multiple connected vehicles at unsignalized intersections. *Transportation Research Part C: Emerging Technologies*, 93:322–334, 2018.

[249] Can Xu, Wanzhong Zhao, Lin Li, Qingyun Chen, Dengming Kuang, and Jianhao Zhou. A nash q-learning based motion decision algorithm with considering interaction to traffic participants. *IEEE Transactions on Vehicular Technology*, 69(11):12621–12634, 2020.

[250] Huile Xu, Shuo Feng, Yi Zhang, and Li Li. A grouping-based cooperative driving strategy for cavs merging problems. *IEEE Transactions on Vehicular Technology*, 68(6):6125–6136, 2019.

[251] Huile Xu, Yi Zhang, Li Li, and Weixia Li. Cooperative driving at unsignalized intersections using tree search. *IEEE Transactions on Intelligent Transportation Systems*, 21(11):4563–4571, 2019.

[252] Li Xu, Jie Hu, Hong Jiang, and Wuqiang Meng. Establishing style-oriented driver models by imitating human driving behaviors. *IEEE Transactions on Intelligent Transportation Systems*, 16(5):2522–2530, 2015.

[253] Linghui Xu, Jia Lu, Bin Ran, Fang Yang, and Jian Zhang. Cooperative merging strategy for connected vehicles at highway on-ramps. *Journal of Transportation Engineering, Part A: Systems*, 145(6):04019022, 2019.

[254] Wei Xu, Hong Chen, Junmin Wang, and Haiyan Zhao. Velocity optimization for braking energy management of in-wheel motor electric vehicles. *IEEE Access*, 7:66410–66422, 2019.

[255] Xin Xu, Lei Zuo, Xin Li, Lilin Qian, Junkai Ren, and Zhenping Sun. A reinforcement learning approach to autonomous decision making of intelligent vehicles on highways. *IEEE Transactions on Systems, Man, and Cybernetics: Systems*, 50(10):3884–3897, 2018.

[256] Zhijiang Xu, Wanzhong Zhao, Chunyan Wang, and Yifan Dai. Local path planning and tracking control of vehicle collision avoidance system. *Transactions of Nanjing University of Aeronautics and Astronautics*, 35(4):729–738, 2018.

[257] Qingwen Xue, Ke Wang, Jian John Lu, and Yujie Liu. Rapid driving style recognition in car-following using machine learning and vehicle trajectory data. *Journal of Advanced Transportation*, 2019, 2019.

[258] Fei Yan, Yi-Sha Liu, and Ji-Zhong Xiao. Path planning in complex 3d environments using a probabilistic roadmap method. *International Journal of Automation and Computing*, 10(6):525–533, 2013.

[259] Da Yang, Shiyu Zheng, Cheng Wen, Peter J Jin, and Bin Ran. A dynamic lane-changing trajectory planning model for automated vehicles. *Transportation Research Part C: Emerging Technologies*, 95:228–247, 2018.

[260] Liu Yang, Rui Ma, H Michael Zhang, Wei Guan, and Shixiong Jiang. Driving behavior recognition using eeg data from a simulated car-following experiment. *Accident Analysis & Prevention*, 116:30–40, 2018.

[261] Peng Yang, Ke Tang, Jose A Lozano, and Xianbin Cao. Path planning for single unmanned aerial vehicle by separately evolving waypoints. *IEEE Transactions on Robotics*, 31(5):1130–1146, 2015.

[262] Shuo Yang, Hongyu Zheng, Junmin Wang, and Abdelkader El Kamel. A personalized human-like lane-changing trajectory planning method for automated driving system. *IEEE Transactions on Vehicular Technology*, 2021.

[263] Xiaoguang Yang, Xiugang Li, and Kun Xue. A new traffic-signal control for modern roundabouts: method and application. *IEEE Transactions on Intelligent Transportation Systems*, 5(4):282–287, 2004.

[264] Chen YongBo, Mei YueSong, Yu JianQiao, Su XiaoLong, and Xu Nuo. Three-dimensional unmanned aerial vehicle path planning using modified wolf pack search algorithm. *Neurocomputing*, 266:445–457, 2017.

[265] Je Hong Yoo and Reza Langari. Stackelberg game based model of highway driving. In *Dynamic Systems and Control Conference*, volume 45295, pages 499–508. American Society of Mechanical Engineers, 2012.

[266] Je Hong Yoo and Reza Langari. A stackelberg game theoretic driver model for merging. In *Dynamic Systems and Control Conference*, volume 56130, page V002T30A003. American Society of Mechanical Engineers, 2013.

[267] Changxi You, Jianbo Lu, Dimitar Filev, and Panagiotis Tsiotras. Highway traffic modeling and decision making for autonomous vehicle using reinforcement learning. In *2018 IEEE Intelligent Vehicles Symposium (IV)*, pages 1227–1232. IEEE, 2018.

[268] Chao Yu, Xin Wang, Xin Xu, Minjie Zhang, Hongwei Ge, Jiankang Ren, Liang Sun, Bingcai Chen, and Guozhen Tan. Distributed multiagent coordinated learning for autonomous driving in highways based on dynamic coordination graphs. *IEEE Transactions on Intelligent Transportation Systems*, 21(2):735–748, 2019.

[269] Chuanyang Yu, Yanggu Zheng, Barys Shyrokau, and Valentin Ivanov. Mpc-based path following design for automated vehicles with rear wheel steering. In *2021 IEEE International Conference on Mechatronics (ICM)*, pages 1–6. IEEE, 2021.

[270] Hai Yu, Finn Tseng, and Ryan McGee. Driving pattern identification for ev range estimation. In *2012 IEEE International Electric Vehicle Conference*, pages 1–7. IEEE, 2012.

[271] Hongtao Yu, H Eric Tseng, and Reza Langari. A human-like game theory-based controller for automatic lane changing. *Transportation Research Part C: Emerging Technologies*, 88:140–158, 2018.

[272] Dongdong Yuan and Yankai Wang. An unmanned vehicle trajectory tracking method based on improved model-free adaptive control algorithm. In *2020 IEEE 9th Data Driven Control and Learning Systems Conference (DDCLS)*, pages 996–1002. IEEE, 2020.

[273] Ming Yue, Lu Yang, Xi-Ming Sun, and Weiguo Xia. Stability control for fwid-evs with supervision mechanism in critical cornering situations. *IEEE Transactions on Vehicular Technology*, 67(11):10387–10397, 2018.

[274] Yao ZeHao, LiQian Wang, Ke Liu, and YuanQing Li. Motion prediction for autonomous vehicles using resnet-based model. In *2021 2nd International Conference on Education, Knowledge and Information Management (ICEKIM)*, pages 323–327. IEEE, 2021.

[275] Wei Zhan, Liting Sun, Di Wang, Haojie Shi, Aubrey Clausse, Maximilian Naumann, Julius Kummerle, Hendrik Konigshof, Christoph Stiller, Arnaud de La Fortelle, et al. Interaction dataset: An international, adversarial and cooperative motion dataset in interactive driving scenarios with semantic maps. *arXiv preprint arXiv:1910.03088*, 2019.

[276] Han Zhang, Bo Heng, and Wanzhong Zhao. Path tracking control for active rear steering vehicles considering driver steering characteristics. *IEEE Access*, 8:98009–98017, 2020.

[277] Han-ye Zhang, Wei-ming Lin, and Ai-xia Chen. Path planning for the mobile robot: A review. *Symmetry*, 10(10):450, 2018.

[278] Kuoran Zhang, Jinxiang Wang, Nan Chen, Mingcong Cao, and Guodong Yin. Design of a cooperative v2v trajectory-planning algorithm for vehicles driven on a winding road with consideration of human drivers characteristics. *IEEE Access*, 7:131135–131147, 2019.

[279] Kuoran Zhang, Jinxiang Wang, Nan Chen, and Guodong Yin. A non-cooperative vehicle-to-vehicle trajectory-planning algorithm with consideration of drivers characteristics. *Proceedings of the Institution of Mechanical Engineers, Part D: Journal of Automobile Engineering*, 233(10):2405–2420, 2019.

[280] Lijun Zhang, Wei Xiao, Zhuang Zhang, and Dejian Meng. Surrounding vehicles motion prediction for risk assessment and motion planning of autonomous vehicle in highway scenarios. *IEEE Access*, 8:209356–209376, 2020.

[281] Linjun Zhang and Eric Tseng. Motion prediction of human-driven vehicles in mixed traffic with connected autonomous vehicles. In *2020 American Control Conference (ACC)*, pages 398–403. IEEE, 2020.

[282] Rui Zhang, Nengchao Lv, Chaozhong Wu, and Xinping Yan. Speed control for automated highway vehicles under road gradient conditions. In *CICTP 2012: Multimodal Transportation Systems Convenient, Safe, Cost-Effective, Efficient*, pages 1327–1336. 2012.

[283] Sheng Zhang and Xiangtao Zhuan. Research on tracking improvement for electric vehicle during a car-following process. In *2020 Chinese Control And Decision Conference (CCDC)*, pages 3261–3266. IEEE, 2020.

[284] Ting Zhang, Wenjie Song, Mengyin Fu, Yi Yang, and Meiling Wang. Vehicle motion prediction at intersections based on the turning intention and prior trajectories model. *IEEE/CAA Journal of Automatica Sinica*, 2021.

[285] Xiaoxue Zhang, Jun Ma, Zilong Cheng, Sunan Huang, Shuzhi Sam Ge, and Tong Heng Lee. Trajectory generation by chance-constrained nonlinear mpc with probabilistic prediction. *IEEE Transactions on Cybernetics*, 51(7):3616–3629, 2021.

[286] Yiran Zhang, Peng Hang, Chao Huang, and Chen Lv. Human-like interactive behavior generation for autonomous vehicles: A bayesian game-theoretic approach with turing test. *Authorea Preprints*, 2021.

[287] Yiwen Zhang, Qingyu Li, Qiyu Kang, and Yuxiang Zhang. Autonomous car motion prediction based on hybrid resnet model. In *2021 International Conference on Communications, Information System and Computer Engineering (CISCE)*, pages 649–652. IEEE, 2021.

[288] J Zhao and W Liu. Study on dynamic routing planning a-star algorithm based on cooperative vehicles infrastructure technology. *J Comput Inf Syst*, 11(12):4283–4292, 2015.

[289] Rui Zheng, Chunming Liu, and Qi Guo. A decision-making method for autonomous vehicles based on simulation and reinforcement learning. In *2013 International Conference on Machine Learning and Cybernetics*, volume 1, pages 362–369. IEEE, 2013.

[290] Yang Zheng, Shengbo Eben Li, Keqiang Li, and Wei Ren. Platooning of connected vehicles with undirected topologies: Robustness analysis and distributed h-infinity controller synthesis. *IEEE Transactions on Intelligent Transportation Systems*, 19(5):1353–1364, 2017.

[291] Bin Zhou, Yunpeng Wang, Guizhen Yu, and Xinkai Wu. A lane-change trajectory model from drivers vision view. *Transportation Research Part C: Emerging Technologies*, 85:609–627, 2017.

[292] Jiacheng Zhu, Shenghao Qin, Wenshuo Wang, and Ding Zhao. Probabilistic trajectory prediction for autonomous vehicles with attentive recurrent neural process. *arXiv preprint arXiv:1910.08102*, 2019.

[293] Qijie Zou, Yingli Hou, and Kang Xiong. An overview of the motion prediction of traffic participants for host vehicle. In *2019 Chinese Control Conference (CCC)*, pages 7872–7877. IEEE, 2019.

Index

For Product Safety Concerns and Information please contact our EU
representative GPSR@taylorandfrancis.com
Taylor & Francis Verlag GmbH, Kaufingerstraße 24, 80331 München, Germany

www.ingramcontent.com/pod-product-compliance
Ingram Content Group UK Ltd.
Pitfield, Milton Keynes, MK11 3LW, UK
UKHW020203200926
475702UK00004B/597
9781032262086